Advances in Industrial Control

Springer
*London
Berlin
Heidelberg
New York
Barcelona
Budapest
Hong Kong
Milan
Paris
Santa Clara
Singapore
Tokyo*

Other titles published in this Series:

Parallel Processing for Jet Engine Control
Haydn A. Thompson

Parallel Processing in Digital Control
D. Fabian, Garcia Nocetti and Peter J Fleming

Intelligent Seam Tracking for Robotic Welding
Nitin Nayak and Asok Ray

Nonlinear Process Control: Applications of Generic Model Control
Edited by Peter L. Lee

Expert Aided Control System Design
Colin Tebbutt

Modeling and Advanced Control for Process Industries, Applications to Paper Making Processes
Ming Rao, Qijun Xia and Yiquan Ying

Robust Multivariable Flight Control
Richard J. Adams, James M. Buffington, Andrew G. Sparks and Siva S. Banda

Modelling and Simulation of Power Generation Plants
A.W. Ordys, A.W. Pike, M.A. Johnson, R.M. Katebi and M.J. Grimble

Model Predictive Control in the Process Industry
E.F. Camacho and C. Bordons

H_∞ Aerospace Control Design: A VSTOL Flight Application
R.A. Hyde

Neural Network Engineering in Dynamic Control Systems
Edited by Kenneth Hunt, George Irwin and Kevin Warwick

Neuro-Control and its Applications
Sigeru Omatu, Marzuki Khalid and Rubiyah Yusof

Energy Efficient Train Control
P.G. Howlett and P.J. Pudney

Marija D. Ilić and Shell Liu

Hierarchical Power Systems Control
Its Value in a Changing Industry

With 38 Figures

Marija D. Ilić, Dr
Massachusetts Institute of Technology
Department of Electrical Engineering and Computer Science
Laboratory for Electromagnetic and Electronic Systems
77 Massachusetts Avenue, Cambridge, Massachusetts 02139, USA

Shell Liu, Dr
Transmission Technology Institute
ABB Power T&D Company Inc
1021 Main Campus Drive, Raleigh, North Carolina 27606, USA

Series Editors
Michael J. Grimble, Professor of Industrial Systems and Director
Michael A. Johnson, Reader in Control Systems and Deputy Director

Industrial Control Centre, Department of Electronic and Electrical Engineering, Graham Hills Building, 60 George Street, Glasgow G1 1QE, UK

ISBN 3-540-76031-8 Springer-Verlag Berlin Heidelberg New York

British Library Cataloguing in Publication Data
A catalogue record for this book is available from the British Library

Library of Congress Cataloging-in-Publication Data
Ilić, Marija D., 1951-
   Hierarchical power systems control : its value in a changing
industry / Marija Ilic and Shell Liu..
   p. cm. - - (Advances in industrial control)
   Includes index.
   ISBN 3-540-76031-8 (pbk. : alk. paper)
   1. Electric power plants - - Management. 2. Electric power systems -
- Control. 3. Electric utilities - - Management. 4. Competition.
   I. Liu, Shell, 1962-   . II. Title. Series.
   TK1191.I43 1996   96-2362
   333.79'23'0117 - - dc20   CIP

Apart from any fair dealing for the purposes of research or private study, or criticism or review, as permitted under the Copyright, Designs and Patents Act 1988, this publication may only be reproduced, stored or transmitted, in any form or by any means, with the prior permission in writing of the publishers, or in the case of reprographic reproduction in accordance with the terms of licences issued by the Copyright Licensing Agency. Enquiries concerning reproduction outside those terms should be sent to the publishers.

© Springer-Verlag London Limited 1996
Printed in Great Britain

The publisher makes no representation, express or implied, with regard to the accuracy of the information contained in this book and cannot accept any legal responsibility or liability for any errors or omissions that may be made.

Typesetting: Camera ready by authors
Printed and bound at the Athenæum Press Ltd., Gateshead, Tyne and Wear
69/3830-543210 Printed on acid-free paper

SERIES EDITORS' FOREWORD

The series Advances in Industrial Control aims to report and encourage technology transfer in control engineering. The rapid development of control technology impacts all areas of the control discipline. New theory, new controllers, actuators, sensors, new industrial processes, computer methods, new applications, new philosophies,......, new challenges. Much of this development work resides in industrial reports, feasibility study papers and the reports of advanced collaborative projects. The series offers an opportunity for researchers to present an extended exposition of such new work in all aspects of industrial control for wider and rapid dissemination.

Power generation and transmission companies are changing in both the US and Europe. De-regulation, privatisation and the introduction of competition to the electricity market are common themes in these areas. Marija Ilic and Shell Liu have produced this book to report their recent research on both the economic and operational impact of these changes. They have been spurred on in their endeavours by significant industrial input and support. As a result, seminal results in the hierarchical and co-ordination methods used to model, control and operate power networks are reported. Also covered in the text are the problems of Secondary Voltage Control using reactive power resources. This problem is also the subject of research in Italy, Spain and France and once again the volume contributes useful and interesting ideas to the subject. The authors show how their co-ordinated hierarchical concepts apply to this problem.

Apart from power engineers, and power system designs the volume will be of interest to large scale systems researchers for impetus and inspiration. There will also be a transfer of some of the ideas to other large scale industrial systems areas; an avenue yet to be explored. The volume is a very welcome addition to the AIC Series.

M.J. Grimble and M.A. Johnson
Industrial Control Centre
Glasgow, Scotland, UK

PREFACE

This monograph is written under rather unusual circumstances. Rarely is theoretical work seen as being of immediate practical need in industry. Technology transfer is slow, if it happens at all. But the situation in the electric power industry is exceptional: At this time, this industry is undergoing striking changes related to deregulation, and little is known about how large electric power systems will function under competition– What are suitable performance objectives, control designs, and economic techniques to keep these large systems working well? These questions can only be answered properly by thorough theoretical analysis.

The authors of this monograph believe that the fundamental questions needing resolution in a changing industry must be viewed in a system- theoretic way in order to benefit society as a whole. They provide here a modeling, analysis, and systems control framework that makes it possible to relate distinctive features of the electric power industry to more conventional supply/demand processes in other industries.

The authors view the network-based electric power industry as a large hierarchical system. The systems control problem then becomes one of keeping the system together in real time within prespecified performance criteria in the face of competitive, profit-driven system inputs. Considerable effort is made to describe which parts of the system must remain coordinated, and which can be distributed, and why.

The industry in transition is not at a technical or economic equilibrium, and strong potential exists for nonjustifiable profit making by informed market participants. The authors' framework makes possible the analysis of closed-loop system dynamics with both technical and economic feedbacks, and provides the basis for nondiscriminatory cost-charging mechanisms for systems control services to competitive market participants. Most important, it simplifies the picture of a very complex system by extracting only relevant information at each level of hierarchy. This is essential for allocating charges to specific market participants in an efficient way: once the performance objectives are standardized, the responsibilities become separable and can be managed in parallel at various levels of the system. If this is not done, it is likely that meeting performance objectives in complex large systems with many independent inputs would be unmanageable and lead to makeshift solutions that would never be fully dependable and nondiscriminatory. In addition, the authors emphasize, without performance-based economic signals for systems control services, the incentives for introducing advanced technology to enhance systemwide performance may be lost.

It is hoped that readers will apply these ideas in industry or pursue them in research. The authors' framework is structure-independent and can be used to evaluate performance under several possible industry structures. The concepts described are evolutionary, not revolutionary, relative to existing systems control principles. As such, they could be tested in practice without the need for many years of new development.

In memory of my dear parents,
Vera and Dragi,
some of whose
strength and stubborness
I was lucky to inherit.

M.D.I.

Dedicated to my dear wife,
Mee,
with gratitude for
her constant unconditional support
and encouragement.

S.X.L.

ACKNOWLEDGEMENTS

The initial research efforts underlying this text were made under the sponsorship of Electricité de France (EDF) as part of an investigation of the need and means for coordinating regional voltage regulation in the French network. We particularly appreciate the early input provided by Christine Vialas and Michael Athans during that project. The basic approach presented here was conceived during that project in the context of studying automatic voltage control (AVC) as a hierarchical control problem. We often studied analogies between our approach and the present automatic generation control (AGC) scheme.

Shortly after the EDF project ended, in 1993, we attended a meeting with several utility people who were concerned about the viability and value of controls under competition. Some worried that under competition there simply would not be enough units participating in systems control to regulate frequency within recommended (not mandatory) bounds. Greg Cucchi of PECO Energy Co. asked how one would define a control area with an independent power producer in the middle of it. Would this be considered a single control area or one within the other, and who would be responsible for regulation? Al DiCaprio of Pensylvania-New Jersey-Maryland (PJM) and Mike Potishnak of the New England Power Exchange (NEPEX) reassured us that the framework we described at this meeting was worth pursuing. The discussions at this meeting strongly motivated us to pursue our investigation of AGC under competition. Our work in this area is continuing, under the sponsorship of the U.S. Department of Energy (DOE), whose support is greatly appreciated.

We also appreciate very much the support of the ad hoc industry group that met with us several times to review our DOE-sponsored work. Art Garfield of Ohio Edison Company and Lou Leffler of Public Gas and Electric Services Company, leaders of the General Agreement on Parallel Paths (GAPP) project, provided outstanding inspiration at these meetings and kept us going. We are grateful for the input provided by Brendan Kirby of Oak Ridge National Laboratory and Charles King of the New York Power Pool.

Another highlight on the way to this monograph was M. Ilić's correspondence with Pravin Varaiya of the University of California at Berkeley, who wanted to see the proof for our assertion that a certain level of coordination is essential for keeping the system together under competition. We thank Pravin for his time and interest.

We also wish to thank two other people, who taught us a lot and who indirectly helped convince us to complete this monograph. They are Les Fink of ECC, Inc. and Ralph Massielo of ABB/SCI, Inc. Our field benefits tremendously from the insight of visionaries like Les and Ralph.

On a more personal note, M. Ilić wishes to thank Ray Dunlop of New England Power Service Company for his encouragement and help in creating her research agenda for power systems at MIT shortly after she arrived from Illinois. Without his strong encouragement, she might have given up by now. Another very special person, from whom she learned a great deal about economics, is Frank Graves of Brattle/IRI, Inc. He asks the hardest questions!

We also appreciate Alice Cheyer's help in improving the writing of this monograph, and the help of Chien-Ning and Yong Yoon with the art work.

Finally, we wish to express our heartfelt thanks to our families at home and our family at MIT. Without their strong support and understanding, this monograph could not have been written.

CONTENTS

1 **Introduction: Basic Assumptions and Concepts**   1
   1.1 Importance of the envisioned control structures in a changing industry   3
   1.2 System regulation issues affected by the vertical separation of the transmission grid from generation   5
   1.3 Organization of this text   6

2 **The Nested Hierarchy as a System Structure in a Changing Industry**   11
   2.1 Principles of existing horizontally structured electric power systems   11
   2.2 Industry changes leading to the nested hierarchy structure . .   14
   2.3 Examples of new industry arrangements as particular cases of the nested hierarchy structure   16
   2.4 The need for new control structures   18
   2.5 Can generation-based regulation be made price-competitive?   19
   2.6 Relevance of dynamic problem formulation over mid- and long-term horizons   20

3 **Performance Criteria Relevant to Operating Interconnected Electric Power Systems**   23
   3.1 Dynamics of system inputs to which the control responds . .   24
   3.2 Time frames for present performance objectives   25
   3.3 Modeling for systems control services in a changing industry .   27
   3.4 Performance criteria at the subsystem level   30
     3.4.1 Static optimization objectives   31

	3.4.2	Dynamic optimization objectives	31
3.5		Static optimization in an open access system	33
	3.5.1	Some assumptions under which present optimal scheduling algorithms are designed at the system level	34
	3.5.2	Generation cost minimization: Ideal technical efficiency	35
	3.5.3	Basic operating cost of keeping the system together	35
	3.5.4	Achievable technical efficiency in the regulated industry	41
	3.5.5	Assumptions that do not hold in a deregulated industry	42
	3.5.6	Need for relaxing the demand-related assumptions (2 and 4)	43
	3.5.7	Need to reconsider the performance objectives (assumptions 1–3)	44
	3.5.8	Hierarchical structures in a distributed industry	45
	3.5.9	Achievable efficiency under open access–ISO market level	46
	3.5.10	Achievable efficiency of competitive supply and demand	46
	3.5.11	Achievable economic efficiency of generation-based systems control	47
	3.5.12	Need for coordinated generation-based systems control in support of competitive markets	47
	3.5.13	Optimal structure for operating and pricing electric power systems under open access	48
3.6		Static optimization of a horizontally structured system	49
3.7		Present criteria for mid- and long-term dynamic performance	51
	3.7.1	Criteria for load frequency control (LFC)/ automatic generation control (AGC)	52
	3.7.2	Dynamic performance objectives over long-term horizons in a horizontally structured industry	54
	3.7.3	Functional requirements for advanced LFC/AGC in a changing industry	54
	3.7.4	Conceptual problems with meeting mid-term dynamic performance objectives by means of present AGC in a changing industry	55
	3.7.5	Conceptual problems with meeting long-term performance objectives in a changing industry	55
3.8		Static performance criteria for reactive power/voltage support	57
	3.8.1	Criteria for mid- and long-term voltage control (AVC) at a subsystem level	57

3.9	Summary		58

4 Structural Modeling and Control Design Using Interaction Variables — 61

- 4.1 Structural modeling 61
  - 4.1.1 Modeling issues 63
  - 4.1.2 Modeling process 66
  - 4.1.3 Local dynamics 67
  - 4.1.4 Network constraints 68
  - 4.1.5 Structural dynamic model 69
  - 4.1.6 Control-induced time scale separation 70
- 4.2 Hierarchical control design 72
  - 4.2.1 Controllability 72
  - 4.2.2 Conventional secondary-level control 73
  - 4.2.3 Improved secondary-level control 74
  - 4.2.4 Quasi-static interaction variables 75
- 4.3 Tertiary level coordination 77
- 4.4 New tertiary-level aggregate model 78
- 4.5 Comparison of the proposed control structures to those used at present 79
- 4.6 Summary 80

5 Generation-Based Regulation of Real Power/Frequency — 83

- 5.1 State of the art and potential problems of frequency regulation — 83
- 5.2 New modeling 84
  - 5.2.1 Local dynamics 84
  - 5.2.2 Network coupling 88
  - 5.2.3 Regional dynamics 90
- 5.3 Analysis 92
  - 5.3.1 Network properties 92
  - 5.3.2 Structural singularity 93
  - 5.3.3 Inter-area dynamics 93
  - 5.3.4 Computation of inter-area variables 94
  - 5.3.5 Interpretation of inter-area variables 97
  - 5.3.6 Comparisons with conventional models 98

	5.3.7	An example	101
5.4		Model derivations	102
	5.4.1	Quasi-static model	102
	5.4.2	Generator power model	108
5.5		Control design	109
	5.5.1	Regulating frequency at the secondary level	109
	5.5.2	Automated regulation of tie-line flows at the tertiary level	113
5.6		Summary	114

6 Generation-Based Regulation of Reactive Power/Voltage — 119

6.1		Modeling	121
	6.1.1	Local dynamics	122
	6.1.2	Network constraints	124
	6.1.3	Structural dynamic model	127
6.2		Quasi-static voltage model	127
6.3		Quasi-static interaction variables	129
6.4		Voltage regulation	133
6.5		Regional voltage control	133
	6.5.1	Conventional secondary-level control	135
	6.5.2	Improved secondary-level control	137
	6.5.3	The 9-bus example	138
6.6		Tertiary coordination	138
	6.6.1	Performance criteria	139
6.7		New tertiary level-aggregate models	141
	6.7.1	Centralized aggregate models	145
	6.7.2	The 9-bus example	146
	6.7.3	Fully centralized optimization	148
	6.7.4	Fully decentralized optimization	151
	6.7.5	Combined centralized/decentralized optimization	152
	6.7.6	Simulations study of the French power network	155
	6.7.7	IAVC	158
	6.7.8	Control at the tertiary level	159
6.8		Summary	159

7 The Value of Generation-Based Regulation: Competition Versus Coordination — 169

7.1 Relevant optimality questions for determining the value of control services 172

7.2 Control-dependent values of subsystems in a competitive environment 173

7.3 Systems control structure-related issues 175

7.4 Long-term stability of decentralized systems control services . 177

7.5 Achievable optimality as a function of the level of control coordination 180

7.6 Limitations of existing systems control in a competitive environment 182

7.7 Proposed approach to real-time systems control and its pricing in a competitive market 183

    7.7.1 Basic steps for linking technical and pricing processes 185

    7.7.2 Operations planning for the anticipated contract at the tertiary level 186

    7.7.3 Estimating at the tertiary level the economic value of systems control services to the contract participants i_1 and i_2 187

    7.7.4 Pricing for systems control at the ISO (tertiary) level to accommodate transaction $i_1 - i_2$ 188

    7.7.5 Contribution of individual components to the economic values at buses i_1 and i_2 188

    7.7.6 Interpretation of our approach in terms of generalized localized marginal costs 190

    7.7.7 From cost-based to value-based future pricing: incentives for high-quality systems control services 191

    7.7.8 Revisiting the "poolco" structure 192

    7.7.9 Revisiting the bilateral structure 192

7.8 Summary 193

8 Network-Based System Regulation — 195

8.1 Engineering issues and opportunities in operating power transmission grids of the future 195

8.2 Recent changes affecting the transmission grid and their relation to the basic engineering issues 196

8.3 A brief review of the present principles for regulating a transmission grid and the power system 198

8.4	The basic planning problem on a transmission grid	200
8.5	Operating problems using mechanically switched reactive devices	205
8.6	Opportunities and problems presented by very fast regulation of the transmission grid: FACTS trends	208
8.7	Direct flow control via FACTS devices	209
	8.7.1 All tie lines directly controlled	210
	8.7.2 Only a subset of tie lines directly controlled	213
8.8	Summary	216

9 Conclusions 219

9.1 Summary of our approach to linking technical and economic processes under competition 221

9.2 Relevance of our proposed modeling and control framework . 223

9.3 A Final Word 224

Bibliography 226

Index 238

CHAPTER 1
INTRODUCTION: BASIC ASSUMPTIONS AND CONCEPTS

Electric power systems in the United States and many other parts of the world are undergoing drastic restructuring. The trend is toward deregulating an industry that has traditionally been a regulated monopoly to allow for economic competition.

Much confusion arises in discussing technical and economic concepts supporting the two extremes of industry structure. At one extreme, some advocate preserving and building upon the present technical framework and standards for operating an interconnected system while yet accommodating competitive non-utility-owned generation. At the other extreme are those who believe electric power systems should be restructured based entirely on economic competition. A change of thinking is needed in both engineering and economic circles to put the debate in a meaningful framework, which is essential for ensuring that, whichever industry structure prevails, basic principles of technical operation as well as the free market economy are met.

This is a tall order for researchers concerned with large dynamic electric power systems. The issues are system-theoretic in their basic nature. True, researchers with a strong background in large systems could identify many new theoretical issues and their solutions within the context of existing problems directly relevant to the power industry. But in order to avoid technical assumptions that may not hold in all cases, one needs a systematic framework for posing these problems.

Much effort has gone over the years into the mathematical modeling of large power systems and, to a lesser extent, systematic control design. Given this, one would think that models for the main operating problems of interest to industry at present would be available. Unfortunately, it turns out that some processes essential for relating technical operation to economics, particularly in a distributed, competitive setting, have never been modeled in a systematic way for power systems. We strongly believe that without such a basic framework for scientific solutions to the problems of power systems under restructuring, not much progress can be made. Thus, we have put considerable effort into establishing a possible starting framework. The essential feature of our approach is that it naturally lends itself to the present structures and could be used to enhance the operation of existing systems. Yet it is generalizable to evolving industry structures as well. Moreover, our approach allows for relating technical processes to economic processes.

Until now, power system monitoring and control have been based on a hierarchical structure where the monitoring and control tasks are shared by different hierarchical levels. Local (primary) controllers on individual generating units are at present decentralized in that they respond to deviations of local outputs from the steady-state set values assigned from higher levels. The steady-state values of primary controllers are determined at a regional (secondary) level, assuming weak interconnections among the regions. The regional controllers are, however, not systematically coordinated at present, leading to deviations from optimal systemwide performance and a possibility of systemwide instability.

The lack of systematic coordination exists in two of the most important control problems of the power system: real power/frequency regulation and reactive power/voltage regulation. In the case of frequency control, a simple coordination scheme known as automatic generation control (AGC) [1]–[5] has been automated throughout the United States and in some other parts of the world. This scheme is based on a reduced information structure that allows for simple automation. Although it has been successful in practice in a relatively static operating environment, hidden problems that may lead to potential loss of systemwide frequency coordination have been identified [7, 8, 6].

In the case of voltage control, systemwide coordination is not automated at present in the United States. Instead, each administratively separate region regulates its own voltages, assuming the effect of changes in neighboring regions negligible. However, as the system experiences unusual reactive power deficiency because of large disturbances, the need for systematic on-line coordination to ensure the security of the interconnected system is becoming evident [9, 10]. Electric power systems in France and Italy are the only systems with automatic regional voltage control [103, 104, 101].

This text introduces a structurally based modeling and control approach for large electric power systems whose interactions among the subsystems are characterized by inter-regional flows. An interconnected system is first decomposed into administratively separate regions. This division makes practical sense because each region has independent controls and makes its own decisions. Next, the dynamics of each region are obtained by combining the local dynamics of individual generator units with algebraic power balance constraints imposed by the transmission network. Throughout most of this work, loads are modeled as disturbances from their expected values. Control design has the main purpose of suppressing the effects of load demand fluctuations from normal operating conditions over a variety of time scales, ranging from seconds through hours.

Introduction of structurally based aggregation is a major contribution of our approach. Structural interaction variables relevant for different control levels are defined, and the corresponding dynamic interaction models are obtained to account for interactions on different time scales among the interconnected regions. A particularly important feature of the interaction variables is that they are interpreted in terms of physically meaningful quantities such as inter-regional power flows. The preservation of the physical meaning of the interaction variables, not done by any other aggregation method at present known in the area of power systems, is critically important for systematic control design aimed at responding to changes in neighboring regions. The derived aggregate models provide a basis for coordinated on-line automatic control of large power systems because they extract only information relevant for each specific control level and are of very low order compared with other models of electric power systems used.

An advantage offered by our approach is that no assumptions with regard to the strength of the interconnections are needed, in contrast to the present state-of-the-art methods, which typically require the weak interconnections assumption. Since the system is decomposed according to its structural properties, the approach is entirely independent of the strength of interconnections. Eliminating the weak interconnection assumption is important because strong interconnections are needed for the inter-regional wheeling imposed in the open access operating mode [12].

Throughout the text, potential use of the framework for developing concepts for competitive electric power systems is described. The link between economic and technical processes is described as the starting formulation for further studies.

1.1 Importance of the envisioned control structures in a changing industry

As the electric power industry restructures, it becomes necessary to provide an assessment of the engineering required to facilitate system operation in new operating modes. The material in this text was written in response to this need.

This text is a result of our continuous research effort toward formulating and solving operating problems in large electric power systems. The fundamental concepts described here evolved from the Ph.D. work of Shell X. Liu under the guidance of Marija Ilić. As a result of this research, a modeling and control framework has been formulated that can be further developed for use under a variety of industry structures.

An example of change in the industry is the much-discussed vertical system disintegration in which transmission is likely to unbundle from generation and distribution into independent administrative units. The disintegration is superposed on already existing horizontally structured utilities [14].

Viewed from a slightly different angle, independent power producers (IPPs) and price-responsive large consumers will become nested within existing defined control areas for system regulation. It is important, also, to take account of the intended process of energy wheeling, in which energy is traded between noncontiguous areas across far electrical distances to accommodate economically attractive supply/demand transactions.

This process requires strong interconnections among the areas and therefore puts in doubt the validity of an assumption often made in the analysis and control of large dynamic power systems, i.e., that the subsystems are weakly interconnected. Just this very fact calls for much new work in the area of hierarchical control structures. Little is known about the conditions under which decentralized control remains stable when the weak interconnection assumption is violated, or about how large the deviations from optimal system performance are when suboptimal control structures are designed at various levels of hierarchically managed systems.

Studying the impact of changes in a particular subsystem within a large interconnected system is generally much harder when the subsystem is part of a nested hierarchy structure than when it is a part of a horizontally structured system.

Our approach to dealing with these industry changes is system-theoretic. We strongly believe that many problems of the changing industry can only be formulated and solved at a system level rather than at an individual component level, and we hope that the material in this text confirms this. Given unprecedented input uncertainties caused by industry restructuring, it is essential to provide a modeling, analysis, and control framework that will keep the system together. This, in turn, requires reviewing the operating and control principles of present system structures and assessing the need for changing them to support new operating modes.

Adequate control structures are important for future operation of large electric power systems because the type of control structures used for keeping the system together in response to strictly profit-driven system inputs (generation and demand) strongly affects systemwide efficiency under competition. While in many other deregulated industries a significant case for economic efficiency can be made by making supply/demand competitive rather than regulated, we show that the real-time operation of electric power systems must remain at least partially centralized. This introduces the need for efficiency evaluation beyond strictly distributed, competitive supply/demand markets and makes the problem distinct from problems that may seem the same in other industries. A recent report summarizing the questions created by deregulation in Europe [15] clearly recognizes the trade-off between competition and coordination in terms of systemwide efficiency.

Viewed from a system-theoretic standpoint, the efficiency problems in electric power systems can be studied as problems of the optimal operation of large hierarchically structured dynamic systems. The difference between the present structures and those most likely to evolve is threefold:

1. Performance objectives are becoming increasingly distributed, i.e., competitive.

2. The horizontal hierarchy structure is evolving into a form of nonstandard nested hierarchy.

3. Administratively separated subsystems are becoming more strongly connected as a result of energy transfers across electrically distant areas.

The approach suggested here is based on the premise that one can define so-called interaction variables at each subsystem level within a nested hierarchy whose dynamics are affected only by changes in other subsystems, not by changes within the subsystem of interest. The notion of such variables was proposed for the first time in [16, 17, 18, 109]. These variables are shown to have a straightforward physical interpretation and can be defined for any hierarchical structure.

1.2 System regulation issues affected by the vertical separation of the transmission grid from generation

Concerning changes in the industry structure related to the potential administrative separation between transmission and generation, it is important to observe that real-time operation mostly relies on generation-based controllers, for a given transmission structure over a long time horizons. This is, strictly speaking, not entirely true, since parameters of the transmission grid actually change in real time, both as a result of unintentional equipment contingencies and in a controlled manner by switching in/out various equipment units such as shunt capacitors, phase-shifting transformers, and series capacitors for voltage regulation purposes. Equipment status also changes for scheduled maintenance.

When attempting to relate processes on the hypothetically separated transmission and generation sides, one must recognize the inherent time-scale separation between slow parametric changes on the transmission side and the considerably faster system state changes on the generation side. Because of this, most equipment changes on the transmission side can be viewed as open-loop over the time horizons of direct interest to generation. The closed-loop system regulation on present systems is mainly done by means of generation-based controls.

Since the concepts described in this text reach into the economics of system operation under competition, we emphasize that the modeling framework in terms of interaction variables naturally lends itself to defining the cost of providing control under competition. The problem of an effective price feedback is then posed using the same basic models. One should note a qualitative distinction between transmission-owned and generation-owned control equipment as known today. Only generation-based controllers have significant operating (fuel) costs. The transmission equipment primarily requires capital cost; once installed, it can be used at practically no operating cost. As the performance/cost trade-off of various future control structures is studied, this should be kept in mind. It turns out that this distinction is essential for the type of pricing schemes one could envision for controls in a competitive supply/demand energy market.

If transmission equipment capable of regulating line flows directly comes into general use, the situation is likely to change. One can view such controllers as electric valves. This type of equipment belongs to the new technology often referred to as flexible AC transmission systems (FACTS) [19, 20]. Once devices of this type become an integral part of the system, we will be able to change transmission grid parameters, in effect, at the same rate as generation-based controllers now act. We recognize this and devote one chapter to fundamental control formulations for this most general situation.

It should be noted that vertical disintegration in this situation may become detrimental to system integrity if the transmission-owned and generation-owned controls are not coordinated in an adequate manner.

1.3 Organization of this text

Chapter 2 summarizes the systemwide control structures at present in use in most large electric power systems. These are referred to in the present industry deregulation discussion as part of ancillary services [21]. The time scales at which these services are currently provided, as well as how and why these services may change in the near future, are discussed. The concept of the nested hierarchy as a necessary structure in a deregulating industry is presented. Examples show how new industry arrangements (existing or proposed) can be viewed as particular cases of the nested hierarchy structure.

We also show why these new arrangements require evolutionary changes in the control structures used at present for frequency and voltage regulation in order to retain acceptable power quality.

Another likely change concerns the size of operating reserves for regulation processes. While in the past an effort was made to operate the system conservatively in case unexpected changes occurred, thereby allowing for significant control margins between the amount of energy actually used and the maximum possible, this is not likely to persist under competition.[1] We will need advanced control structures capable of making the most out of available resources. System operation will need to be more flexible in response to supply/demand in real time for both technical and economic efficiency. The requirements to provide different (non-uniform) power quality to different system participants with adequate control structures is likely to grow in a price-driven environment.

Chapter 3 reviews the performance criteria used at present in regulating electric power systems. The first part is concerned with the type of criteria used for static optimization to meet anticipated demand within operating constraints. All performance criteria of this type are cost-based and fall short of optimizing demand (customer) benefits. We discuss the similarities and differences between present operating criteria and the criteria required for an economically efficient industry in a free market and indicate the possible missing links between the two. The second part of this chapter describes existing performance criteria for system regulation in response to unanticipated slow load deviations, and projects how are they likely to change in a deregulated industry.

In Chapter 4, interaction variables are introduced as the variables whose dynamics are basic for understanding the processes among the subsystems within a large system whose structure is nested. General principles for deriving models using these variables are described. Control problems at various levels of a system and various time horizons are defined in terms of the interaction variables and variables known to the level of interest.

In Chapter 5, we demonstrate the use of interaction variables for reviewing the principles of present real power generation over various time horizons and at various levels of system hierarchy. The recognition of the time scale separation of generation control tasks at different levels of system hierarchy allows for a unified approach to generation control, ranging from seconds to hours in response to uncontrolled system input dynamics (competitive supply/demand). Dynamic economic dispatch is viewed here as an integral part of system regulation over a longer time horizon than in generation-based regulation for frequency quality. This unified formulation of generation-based system regulation is important for regulating the system in response to market-driven system input changes. We indicate in this chapter how one could meet non-uniform performance criteria by adequate control

---

[1] An analogy in other industries is that inventories will be significantly reduced [22].

structures. Particular design is envisioned to accommodate distributed regulation at the subsystem levels, while requiring only minimal regulation at the interconnected system level.

In Chapter 6, the general modeling framework introduced in Chapter 4 is used to pose generation-based reactive power/voltage regulation of an arbitrary industry structure. Hierarchical voltage control structures now implemented in France and Italy are described as illustrations of a more general control structure. We also describe how this structure could be used in some proposed new industry structures. Particular emphasis in this chapter is put on the notion of voltage optimality in large electric power systems. This is a difficult problem, and only partial answers are suggested, but the effectiveness of a voltage control structure will strongly depend on the type of performance criterion selected to reflect deviations from the optimal voltage.

The inefficiencies of competitive supply/demand transactions across electrically distant areas are particularly seen when providing reactive power generation needed to compensate for reactive power losses. It is known that reactive power losses are an order of magnitude higher in typical power systems than real power losses. For this reason, reactive power loss compensation is more efficient when provided at a local level rather than at a systemwide level. The modeling framework used here shows this. We also discuss performance criteria relevant at the systemwide (centralized) level versus performance criteria that can be met at the subsystem level and the implications of this result for suitable control structures for voltage regulation.

Improvement of efficiency under competition is studied in Chapter 7 as a problem of suboptimal performance of decentralized control structures relative to optimal centralized performance. Exact measures of deviations from optimal performance caused by distributed (competitive) control architectures are formulated for an arbitrary industry structure. Next, analysis is provided to illustrate how the efficiency achieved for given competitive supply/demand strongly depends on the specific industry structure, the electrical properties of the interconnected system, the strength of the connections among the subsystems, and the control architecture supporting the process. Based on this analysis, a framework that can provide a pragmatic means of maintaining reliable system operations while simultaneously fostering a competitive supply/demand market is introduced.

While our emphasis up to this point has been on generation-based control structures, Chapter 8 focuses on the transmission network and on recently developed transmission control hardware, including FACTS devices, that allows for direct control of line flows. The chapter is concerned with control problem formulation assuming vertically integrated transmission and generation, and both transmission- and generation-based controls. Caution is suggested because the necessary theory for using both control technologies simultaneously has not been fully developed. We also look at the use of both kinds of controls in an industry structure where the transmission network and generators belong to different owners. Very little is known about effective control structures when this is the case.

Chapter 9 concludes by emphasizing that system operations is more than a series of static optimizations. Operations requires a firm and known base of system resources for its activities. This base is currently provided by operations planners. It ensures the availability of sufficient capacity to respond to disturbances, sufficient operating bands (between minimum and maximum units' operating limits) to respond to expected load changes, and sufficient responsive generation to respond to short-term load generation mismatches. Our proposed framework provides this same base for reliability in the context of an unbundled open access environment. To effect this reliable competitive system the framework differentiates between a decentralized competitive profit-driven market (CM) and a centrally directed services market (SM).

The role of the hierarchical power systems control concepts described here as essential for effective creation and operations of the services market is briefly summarized. It is suggested how the proposed framework employs incentives to integrate the two markets into a *reliable* free-market.

CHAPTER 2
THE NESTED HIERARCHY AS A SYSTEM STRUCTURE IN A CHANGING INDUSTRY

Before introducing the notion of nested hierarchies, considered here as necessary for formulating operation and control in the changing electric power industry, we briefly describe the horizontal hierarchy structure typically found in existing power systems. Next, we introduce the concept of the nested hierarchy and its relation to the changes required in present system operation and control in order to accommodate competitive supply/demand energy transactions while still keeping the system together. Examples of some recently evolved industry structures are shown to be particular cases of the general nested hierarchy.

2.1 Principles of existing horizontally structured electric power systems

Very large electric power systems are currently operated and regulated in a mode characterized by a certain degree of autonomy at each level of hierarchy yet allowing some coordination among the subsystems when necessary. The subsystems are usually defined according to administrative divisions among particular utilities. As a result, each utility enjoys operating autonomy and exercises its ability to coordinate operations when desired for technical or economic reasons.

The typical horizontal arrangement of a very large electric power system is shown in Figure 2.1. The interconnected system may represent a typical power pool consisting of several utilities or an even larger system like the entire eastern United States or western United States.

Within this larger system all power pools are operated at the same level of hierarchy and form the highest level of subsystems within the horizontally arranged hierarchical levels. The lower level of hierarchy is formed by utilities within each power pool. Each utility at present has sufficient generation to meet anticipated load in its area (native load) autonomously when generation scheduling is done according to its operating cost. Out-of-merit generation dispatch within each utility is only exercised when operating constraints demand it. Operating the system in wheeling mode, which requires power transfer to noncontiguous subsystems for economic reasons, is not typical. Only neighboring subsystems usually exchange power by scheduling their generation to meet the agreed-upon tie-line flow exchange and their native loads, and the decision to exchange power is made at this lower level.

Only very few exceptions can be found in which the tie-line flow exchange among subsystems is coordinated at the highest level of hierarchy for economic reasons. The Pennsylvania-New Jersey-Maryland (PJM) pool is one such exception. The New England Power Pool (NEPOOL) is another, although its accounting is done after rather than before the power transfer.

The term *horizontal* is used here to emphasize the fact that all generation units within an administratively defined subsystem are owned by this subsystem and are under its control for generation scheduling in response to native load variation and, less frequently, in response to intentional tie-line flow exchange with neighbors. In NEPOOL the interpretation of such an arrangement is that all units are pool members and can be scheduled by higher levels to achieve most efficient and reliable system operation as the native load deviates from the anticipated, as tie-line flow exchange with neighboring subsystems deviates from anticipated, and (very important) as unexpected equipment (both generation and transmission) outages occur.

Over the years, such horizontally structured systems have developed sets of operating rules concerning system performance in normal and emergency operating conditions. Given the complexity of these systems, the operating philosophy adopted uniformly throughout the world has been one of preventive rather than real-time control. In particular, all subsystems have been operated in such a way that if any single unexpected contingency occurs, the whole system will remain within acceptable operating constraints. This means, in effect, that conservative bounds on control equipment are established to ensure uninterrupted system operation when only a single unexpected discrete event occurs. This operating practice is called the $(n-1)$ security criterion [25]. For strictly technical reasons, discussed later in this text, the operating practice provides sufficient power reserve to accommodate unexpected events at each subsystem level. It can be shown that the larger the subsystem, the less such reserve is needed. Therein lies the efficiency of an interconnected system.

For what follows one should note that the level of reserve, in terms of the

Chapter 2

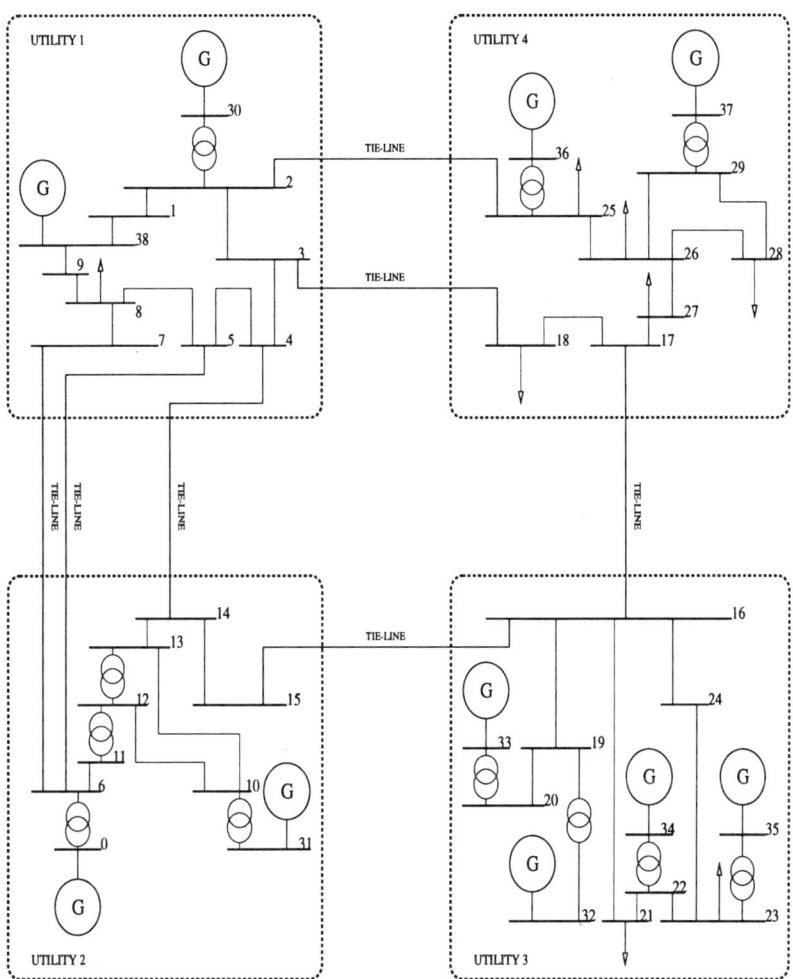

G: UTILITY-OWNED GENERATORS

Figure 2.1: Typical horizontal structure of existing electric power systems

($n - 1$) security criterion, depends on conditions: for peak load, reserve is generally lower than for low load.

2.2 Industry changes leading to the nested hierarchy structure

At present, most of the subsystems in a horizontally structured industry are characterized by generation excess. This is the starting ground for competition among generators for customers, and for buyers to seek the most beneficial suppliers.

In this situation, generation that does not participate in systemwide regulation is seen as driven by prices and does not lend itself to conventional scheduling of centralized economic dispatch at a subsystem level.

Also, the variations in what used to be called native load are hard to predict, since they may become more dynamic than in the past. This is particularly true if retail wheeling, which allows customers to purchase energy in a price-competitive manner outside the subsystem in which they are physically located, becomes a reality.

In addition, generation designated for systemwide regulation may not be owned by the subsystem. If, for example, independent power producers (IPPs) want to participate in systemwide regulation they will have to be incorporated into the existing control architectures in a nondiscriminatory manner rather than pay for this service.

Given dynamic, profit-driven variations in generation and demand in the subsystems, it is likely that the burden of keeping the system together will increase appreciably. This will require truly adaptive regulation schemes.

The fact that, within an administratively defined subsystem such as a utility or a pool, not all generation is owned by and under the jurisdiction of the subsystem calls for a change in structure. This is true of generation intended to meet the native load of the subsystem as well as of generation needed to participate in regulating the subsystem when system inputs deviate from those anticipated.

This changed structure of administratively separated units can be viewed as a nested rather than as a horizontal hierarchy. This nested hierarchy structure is shown in Figure 2.2. A generation unit (I) operated or owned by someone other than a subsystem in a horizontally structured environment is nested inside the subsystem. Similarly, a market-price-modulated demand unit is nested within the subsystem and can no longer be considered to be native to this subsystem in the sense that the subsystem can anticipate its level of demand with certainty.

Chapter 2

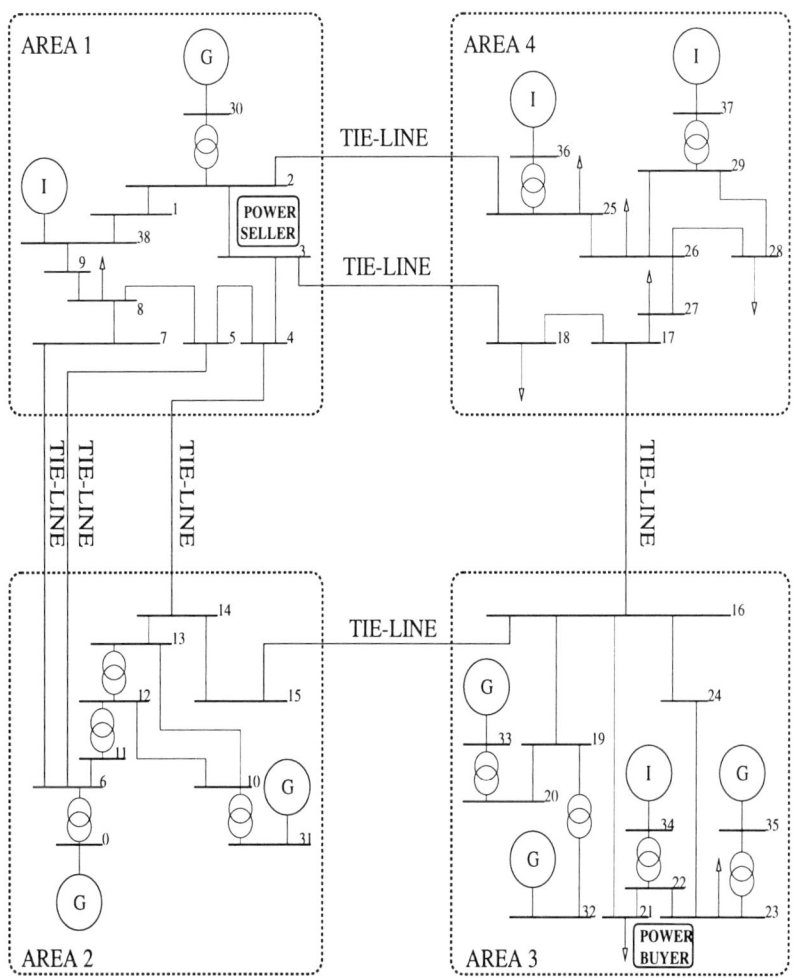

G: UTILITY-OWNED GENERATORS
I: INDEPENDENT POWER PRODUCERS

Figure 2.2: The nested hierarchy structure

2.3 Examples of new industry arrangements as particular cases of the nested hierarchy structure

Most of the analysis in this text is in the context of four representative administrative scenarios:

1. The system is a single administrative unit.

2. The system is horizontally divided into two or more administratively different subsystems (control areas).

3. The system is horizontally divided into two or more administratively different areas, and within each area are nested generating units that belong to independent owners.

4. Fully distributed industry structure.

Structures 1–3 are schematically shown in Figure 2.3.

The first two cases are representative of present energy management, structure 1 corresponding to an electrically isolated power system and structure 2 to a pool of electrically interconnected utilities with individual objectives. Structure 3 would be typical of competitive nested hierarchies consisting of interconnected utilities and non-utility-owned generators (NUGs) within the same area. Energy consumers (loads) traditionally buying power from the area (utility) to which they are directly connected are likely to compete for the least expensive generation across the utility borders. The fourth case assumes a fully distributed supply/demand structure.

In principle, one needs to differentiate between strictly profit-driven supply/demand changes in a competitive market (CM) and coordinated generation/demand control necessary to maintain systemwide performance in response to the actions of individual CM participants. These necessary systems control functions over various time horizons can be viewed as taking place in a services market (SM) [21, 59].

In discussing relevant performance criteria for energy management, two kinds of issues arise: determining performance criteria for an electrically isolated power system, e.g., structure 1, and choosing performance criteria for administratively non-uniform industry structures, e.g., structure 3, where systemwide objectives typically differ from the objectives of individual units.

A "poolco" structure [26, 27], resembling structure 1, involves generation scheduling by a central operator. A bilateral structure, resembling structure 2 or 3, features a decentralized scheme where energy transactions are effected between individual buyers and sellers [28]. A hybrid structure [35, 59] is generally a mix of coordination needed to keep the system together in response

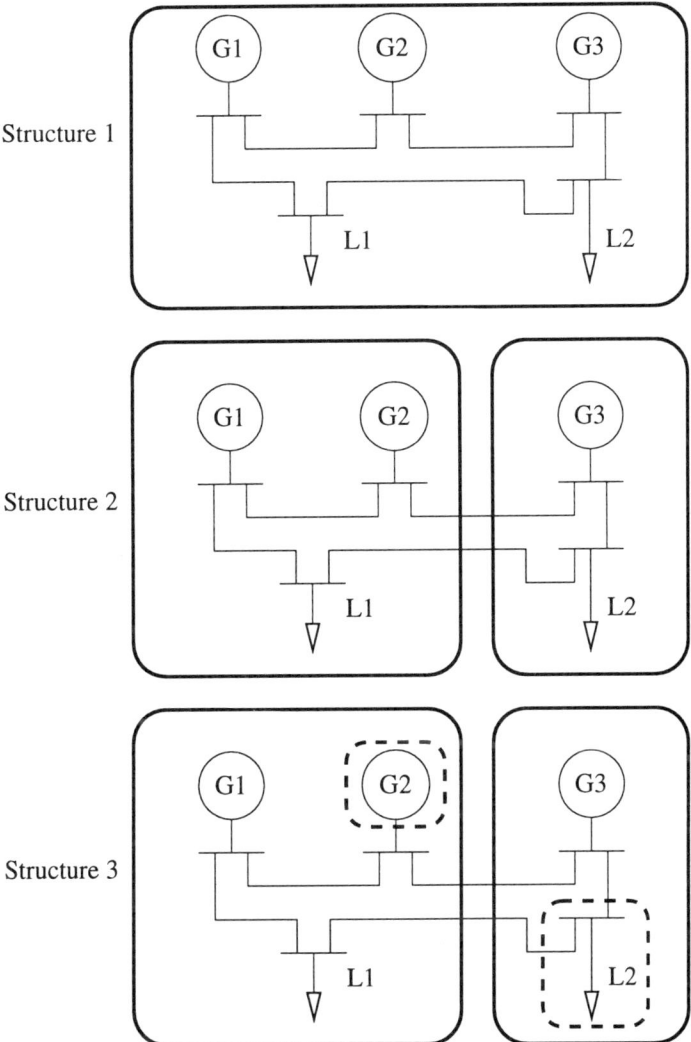

Figure 2.3: Three representative system structures

to decentralized profit-driven system inputs. It generally falls under the category of structure 3. Managing a "poolco" is similar to managing planning and control procedures in current "tight" power pools, whereas a bilateral-based system must resolve the extent to which the individually scheduled transactions can be separated from the grid management itself. Other issues in the bilateral environment include the methodology used by the coordinator of system services, for instance, an independent system operator (ISO) [29]), to prevent or relieve flows that constrain individually scheduled transactions, and the allocation of responsibility for system services (e.g., network protection, constraint relief) among those resources that cause and those that benefit from the conditions requiring those services.

2.4 The need for new control structures

One must recognize that unless new monitoring and control structures are developed to account for the impact of competitive supply/demand activities within each subsystem, it is not going to be possible to meet the basic obligation of keeping the subsystems and the interconnected system functioning within expected performance.

A look at the control structures for real-power generation needed to meet anticipated loads and at dynamic regulation structures such as automatic generation control (AGC) reveals that these schemes rely on very reduced information, which makes controls such as AGC feasible in real time. We show later in this text that this information structure is effective only under certain assumptions, which are not likely to be satisfied in a deregulated industry.

In what follows we show that generally more information and measurement are necessary to incorporate non-utility-owned participants into systemwide regulation. However, it is essential not to introduce a revolutionary change without considering the information and measurement structure on which present operation heavily depends.

The emergence of price-driven generation and, to a lesser extent, price-driven demand calls for competitive, distributed generation. A control structure that would be fully distributed on the generation side is not available at present, not even in concept. Until new control structures capable of serving a competitive energy market in a truly distributed manner are developed, generation-based control structures can only change relative to their present structures.[1]

It is also becoming necessary to allow for new types of system regulation to accommodate non-uniform power quality desired at particular levels of the nested hierarchy. In other words, it is no longer sufficient to regulate

---

[1] It is not obvious to these authors that entirely distributed energy management would be a feasible concept. However, the claim is hard to prove.

frequency and voltage to the same level of quality throughout the entire subsystem. While it has been long recognized that not all customers need the same power quality, very little has been done on providing such service. In a competitive world, in which service is provided at the price, it will become necessary to design control structures capable of providing energy of prespecified quality. In facing this challenge we mention that most of the present regulation schemes, such as AGC, actually do not meet prespecified dynamic performance even at the subsystem or interconnected system level. This is described in detail in Chapters 5 and 6, and it is only mentioned here to make the point that meeting a requirement to provide non-uniform power quality to specific levels in the nested hierarchy structure is a serious technological challenge if one starts from the present state of system regulation. One can view this problem as one of allowing specific market participants to operate at the levels of reliability, frequency and voltage quality, and operating reserve chosen by each of them locally. The regulation problem at the interconnected system level then becomes one of regulating the energy flows among the participants in such a way that the interconnected system remains functional.

2.5 Can generation-based regulation be made price-competitive?

The remaining piece of the puzzle concerned with generation-based controls serving competitive markets is whether and how generation-based controls can be made price-competitive, like generation that does not participate in system regulation. This is a hard task, since the case is made here that at least partial coordination is necessary to regulate system frequency in electric power systems; otherwise, it is impossible to maintain system frequency at 50 Hz or 60 Hz [32]. Solving the puzzle requires full understanding of the real-time interplay between market-driven processes and the system regulation structure being implemented. One must, at least in principle, differentiate between two types of generation-based activities on the system:

- If the transactions are entirely bilateral, and their amounts are not known to the centralized authority, i.e., the operation is entirely decentralized within a given nested hierarchy structure, closed-loop generation controls would be dominant in compensating for systemwide disturbances created by such transactions. This compensatory activity would be over and above any necessary compensation for random demand fluctuation in native load (or what is left of it). Depending on how active the market is, the amount of generation required for this purpose may become much larger than now used for AGC. As currently implemented, AGC is only intended to compensate for deviations in native load.

While from a purely technical standpoint this solution can be shown to be feasible, serious problems may arise in pricing this service in a nondiscriminatory manner, because it becomes increasingly difficult to separate the effect of fluctuations in native load from fluctuations caused by intentional bilateral transactions.

- If the transactions are coordinated at the centralized level by a mechanism such as a futures market [33], it becomes much easier to identify the portion of generation-based controls responding to the active energy market. In particular, at the beginning of each time period when the futures market acts, say each half hour or hour, controlled generation can be scheduled to compensate for losses created by bilateral transactions. It is fairly straightforward to account for loss compensation to each particular transaction and moreover to define an ex ante price (at the beginning of the time period) for the generation-based service. In this case, closed-loop system regulation becomes essential only in between the regular time samples and is responsible for responding to fluctuations in both native load and in transactions that diverge from those prespecified at the beginning of each time period. In this scenario the amount of generation for closed-loop system regulation is significantly smaller than when the mechanism of the futures market is not used. The problem of accounting and pricing in a competitive manner for this service is again a difficult one, since it is not simple to separate random fluctuations in native load from the noncompliance with the market transactions. At least one interesting approach to solving this problem has been offered in [34].

2.6 Relevance of dynamic problem formulation over mid- and long-term horizons

For generation-based controls to remain price-competitive with the generation not committed to systemwide performance, it is important to recognize the following. While in the past the practice has been to use primarily static optimization tools for generation scheduling, such as constrained economic dispatch and optimal power flow (OPF) techniques, for meeting anticipated demand on a daily basis, much greater efficiency can be achieved when optimizing over the longer time horizons determined by the dynamics of competitive markets. This observation is essential for efficient use of units that are already on line as well as for scheduling which units to switch in and out of service. In other words, dynamic programming techniques [30] for optimal performance of generation-based controls in response to a competitive market will become powerful tools for remaining price-competitive. While tools of this type are highly desirable, they are also hard to implement because they require some projections about the dynamics of profit-driven system inputs

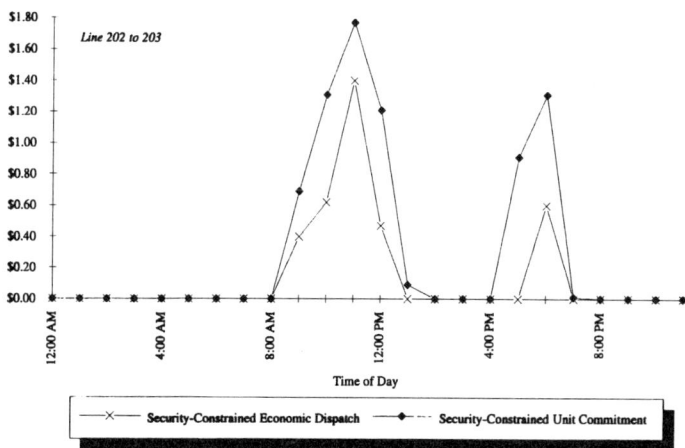

Figure 2.4: Operating cost comparison for meeting demand when generation scheduling is done using OPF and unit commitment (Source: Alphatech, Inc.)

(supply and demand). This requires techniques for projecting the market profile, something that is actively done in many competitive industries. The power industry, whose main role in the future appears to be one of providing operating and control structures in response to a competitive energy market, should develop such techniques in order to remain competitive. An example, familiar to the power industry, that illustrates this need is shown in Figure 2.4. The figure compares operating costs when scheduling is done using OPF versus when unit commitment (i.e., dynamic programming) is used to meet demand over a longer time horizon [31].

While dynamic optimization over longer time periods is known to be more efficient than static optimization, dynamic programming techniques are not routinely used by engineers who operate existing large power systems, primarily because of computational complexity and the huge amount of data required for a typical large system. One possible step for more practical use of such techniques is suggested in this text in Chapter 7. One needs to apply dynamic programming only to minimum-order models relevant for the problem of interest. For example, to schedule energy exchanges among particular levels of the nested hierarchy, it is sufficient to use a model that relates line-flow dynamics among the participants and the generation-based control designated to perform this function. Even on a very large system the order of such a model, when carefully created, may be very low relative to the full model. The approach developed in this text provides a general modeling framework for such purposes. While it is hard to prove that the information in the proposed framework is truly the minimum required for relating processes of direct interest in a hierarchically organized large electric power system, the framework relies on notions already implemented in present control structures. In this sense, it is only an evolution of existing controls and is not revolutionary. The evolution is in terms of accommodating nested, instead of only horizontal, hierarchies.

The proposed framework relies on the notion of interaction variables associated with each particular subsystem within a nested hierarchy. They are defined as variables that are only affected by controls external to the subsystem, and they are fundamental to capturing the interplay among subsystems at various levels of hierarchy. These variables are shown to be closely related to the area control error (ACE) signals that are used at present for AGC. The ACE could be viewed with respect to an existing horizontally organized control area as the simplest known (i.e., minimum) information possible since it is a single (scalar) measurement per control area. It is shown that these interaction variables [17] simply reflect the total supply/demand mismatch at a subsystem level of interest. In this sense, the ACE is shown to be a particular case of an interaction variable in a horizontal hierarchy that under stationary operation reflects the area-level power mismatch [1, 4].

CHAPTER 3
PERFORMANCE CRITERIA RELEVANT TO OPERATING INTERCONNECTED ELECTRIC POWER SYSTEMS

Before addressing the role of coordinated systems control and its performance objectives in a competitive power market, it is essential to appreciate the nature and limitations of the currently available planning and operating tools and procedures. Power systems are managed according to a variety of time frames. These time frames relate to control objectives that the systems must meet in response to independent input changes [36], [59].

To set the stage for understanding the objectives of systems control, we briefly review the types of input changes likely to be seen in a changing industry. It is important to keep in mind that the present system has been designed for qualitatively different types of inputs than, for instance, the market-driven demand and generation. On the other hand, unless otherwise decided, the technical performance objectives are likely to remain unchanged, and they can be introduced in the context of both present and changing operations. As such, they must be standardized and fully understood by the people responsible for introducing price-charging mechanisms for systems control. In this sense, once agreed on, the performance objectives could serve as a good common denominator for comparing performance quality under particular industry structures. If, on the contrary, these are not well defined, the objectives of systems control support under competition will not be well understood, which could lead to potential conflicts of interest among various parties.

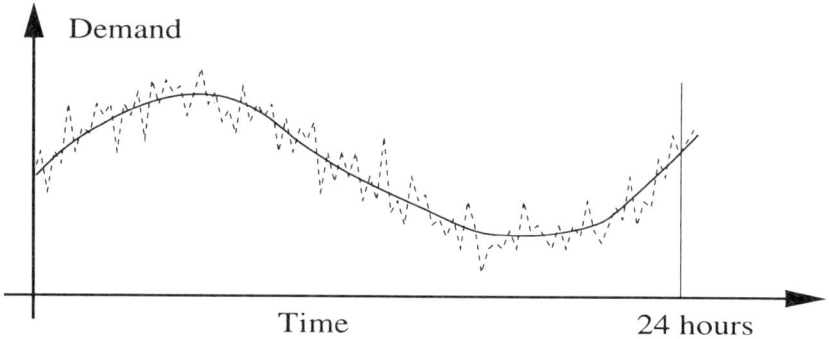
Figure 3.1: Typical diurnal variations in demand

3.1 Dynamics of system inputs to which the control responds

At present, systems control services under normal conditions respond to anticipated demand and its random, small deviations. A typical demand curve for which the scheduling is done is shown in Figure 3.1. This curve represents an aggregate demand as seen at the extra high voltage (EHV) transmission network level.[1] The fluctuations in individual loads are seen in this model as small random fluctuations around this aggregate representation.

In attempting to provide systematic algorithms for system services needed to keep the system together under competition, generally referred to as ancillary services [21], it may be helpful to recognize three qualitatively different causes of system input changes to which ancillaries respond:

1. Firm, long(er) term contracts.

2. Short-term (half hour or less), non-firm contracts.

3. Noncompliance-related changes that reflect deviations from 1 or 2.

In addition we recognize residential load, whose changes are fairly well understood as a function of time.[2]

System inputs of types 1 and 2 are shown in Figure 3.2. The noncompliance may display wide ranges of rate and magnitude at which it evolves. The qualitative difference between 1 and 2 does not arise from their different relative durations but is due to the fact that planning can be done with high confidence only for type 1. Short-term contracts are not known over the time horizons relevant for planning system support. Furthermore, the qualitative

---

[1] Most of the material in this text is related to the EHV transmission level. At a distribution level the relative dynamics of system inputs may be qualitatively different, including the fact that no aggregate load model is used.

[2] Retail wheeling is not accounted for here [21].

Chapter 3

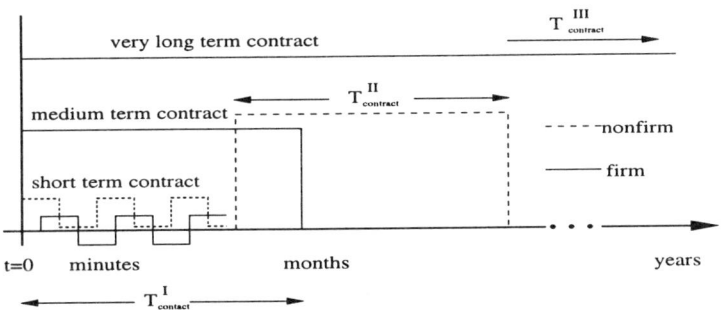

Figure 3.2: Typical firm and short-term, non-firm contracts

difference between 1 and 2, on the one hand, and 3, on the other, comes from the fact that 3 is always an uncertain system disturbance and must be compensated for in real time by automatic control schemes such as AGC. Much of generation-based regulation can be done in a static open-loop manner for firmly scheduled transactions. Participants in a competitive market would generally be concerned with longer-term issues (monthly, daily, hourly contracts), leaving the shorter-term issues (load frequency control, congestion control, disturbance response) to be addressed by the coordinator of system services (ancillaries), such as an independent system operator (ISO). This distinction in time frames affects what system information is useful to which market. *To make this analysis useful in the context of a changing industry, it helps to think of participants in a competitive market as changes in positive/negative demand of types 1 ,2, and 3, to which control responds in order to meet performance criteria of interest.*

The performance objectives described in this chapter should be understood in the context of control objectives to regulate system frequency and voltage within acceptable operating constraints despite changes imposed by these inputs. Furthermore, given this classification, the methods charging for control costs should be systematic and nondiscriminatory.

3.2 Time frames for present performance objectives

At the shortest time frame associated with system operations (less than several seconds) responses are necessarily natural (e.g., generator governor response to frequency deviations). These "natural responses" can be viewed as a community service since they reflect a generator's response to changes in the interconnected system [24]. Response over a period of 0.5–10 minutes to a controller's own mismatch between generation and demand is provided by the regulating units. These are units that can automatically provide a predefined amount of energy within a 10-minute period. Both of the previous functions are activated in response to actual conditions. Economic dispatch is the function that generally incorporates some measure of anticipation of changes. It is used to coordinate the characteristics of the generating resources with the characteristics of the load trend. This function evolves over a 5–30-minute period.

All three of the preceding functions deal with units that are operating. In the present system the coordinator must also project which units should operate in subsequent periods. Unit commitment programs are used to fulfill this function. Operating capacity additions must be planned in advance given the fact that it may take from 1 to 10 hours to restart an available steam generator. Other conditions, such as hydro storage facilities, may require that weekly concerns be included (e.g., start the week with full storage, end the week with no storage, and pump the storage full over the weekend).

The previous functions are related to operations during normal conditions. Provisions for the unexpected must also be made. Sufficient standby capacity must be available to respond to the loss of the single largest contingency. This is provided by operating reserve capacity, which must be available within 10 minutes.

These standard systems control functions are summarized as follows:

Normal State

Time Frame	Function
2–3 seconds	Inertia, loads (not in operations time frame)
7–10 seconds	Governors
0.5–10 minutes	Regulation
5–30 minutes	Economic dispatch, manual control
1–10 hours	Unit commitment, restarting/shutting off units

Emergency State

Time Frame	Function
0–10 minutes	Operating reserve

How much, if anything, of these functions must be provided? The magnitude of operating reserve is currently required by the North American Electric Reliability Council (NERC); this system service must be provided [36]. However, the need for the other functions currently used in control is not specified. NERC specifies a required control performance over a 10-minute period. The need for any of the operating control functions is an empirical matter that is related to the variability of a given demand, the stability of a given supply, and the predictability of both.

We believe that all performance objectives must be standardized and well defined at each level of an otherwise arbitrarily designed nested hierarchy of participants in the energy market. Meeting these objectives should be required in a changing industry that rests on competition and not on voluntary cooperation. Severe penalties must be imposed on participants committed to meeting these objectives when they fail to do so.

In the remainder of this chapter we review present recommended performance objectives at both the subsystem and the interconnected system level. Possible problems with meeting these objectives by means of present systems control schemes in a competitive environment are suggested.

3.3 Modeling for systems control services in a changing industry

The operating philosophy of systems control services under competition remains based on the same principles as at present. It is the one of least-cost operation for acceptable technical specifications.

Because of the complexity involved in defining technical specifications, they should be taken as given for a certain transition period. For the longer-term task of achieving efficient systemwide energy management under competition, technical specifications at the systemwide level may need to be redefined. An immediate study of relaxing the $(n-1)$ security criterion and the implications of this for efficiency should be undertaken. This will allow for exploiting the potential of control tools capable of fast response to system changes, such as pumped storage generation. The economic value of these technologies is significantly lower than it would be if the conservative $(n-1)$ security criterion were relaxed.

A schematic representation of all major systems control functions is shown in Figure 3.3 [76]. The figure indicates the typical temporal and spatial separation of systems control services in a horizontally structured interconnected system.

At present, these functions are primarily driven by changes in demand, as indicated in Figure 3.1. Given this curve, the general model of load at any bus j can be thought of as

$$P_{Lj} = P_{Lj}(t) + P_{Lj}[kT_s] + P_{Lj}[KT_t] \qquad j \in N_C \tag{3.1}$$

Similarly, any generation changes, either in response to load changes or independent of them, have temporally separated components evolving at the fastest random time t, at the slower time order of minutes kT_s and components very slowly evolving on a half hour to an hour basis KT_t, and are represented mathematically as

$$P_{Gi} = P_{Gi}(t) + P_{Gi}[kT_s] + P_{Gi}[KT_t] \qquad i \in N_A \tag{3.2}$$

The reactive power changes are represented similarly as

$$Q_{Lj} = Q_{Lj}(t) + Q_{Lj}[kT_s] + Q_{Lj}[KT_t] \qquad j \in N_C \tag{3.3}$$

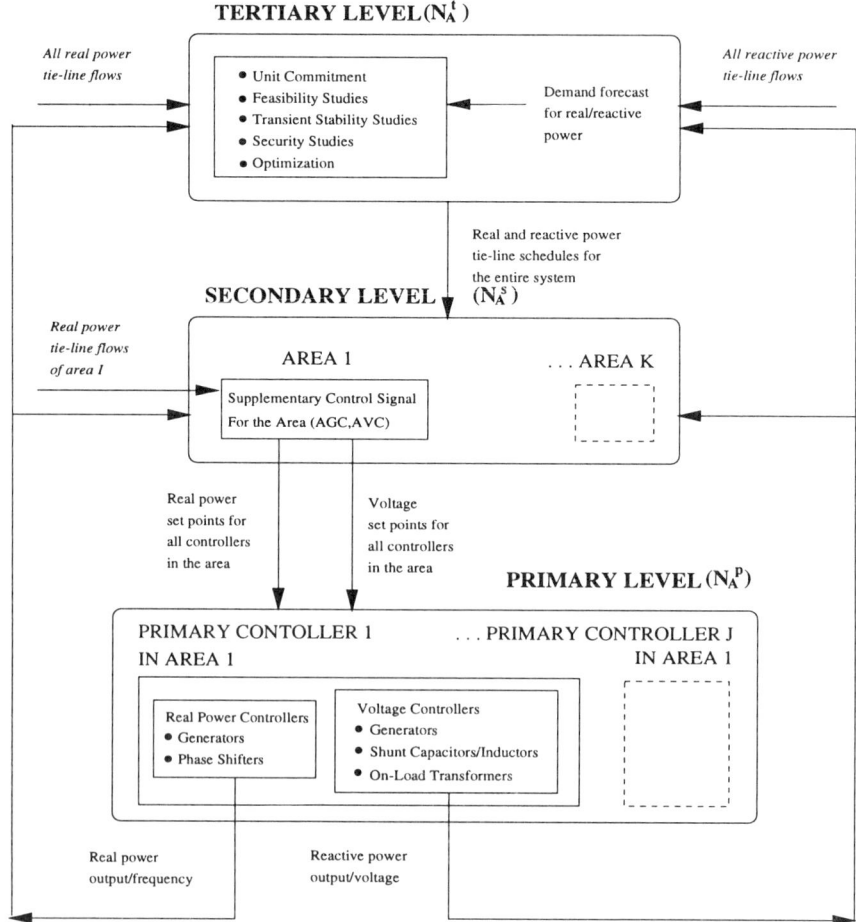

Figure 3.3: Hierarchical measurement/control structure of ancillary services

$$Q_{Gi} = Q_{Gi}(t) + Q_{Gi}[kT_s] + Q_{Gi}[KT_t] \quad i \in N_A \quad (3.4)$$

Sets N_A and N_C denote, respectively, market members committed to (ancillary) systems control services and strictly profit-driven competitive market participants (CMPs) [59]. Ancillary control actions adjust real and reactive power generation outputs P_{Gi} and Q_{Gi}, respectively, in response to deviations in real and reactive power at the competitive supply/demand system inputs.[3] Their temporal separation is indicated in Figure 3.3, showing

- Automatic primary controllers for fast stabilization in response to fast input deviations. Representatives of systems control services of this type are excitation systems, power system stabilizers, static Var compensators (set N_A^p in Figure 3.3).

- Automatic generation scheduling for frequency and voltage quality at each area level, performed at the rate kT_s.[4] This is a centralized function at each control area level. The technical concept and implementation of automatic generation control (AGC) are truly ingenious, and as a result they have worked extremely well (set N_A^s in Figure 3.3).

- Constrained economic dispatch at the rate KT_t (set N_A^t in Figure 3.3).

The fastest controls are called primary controls. At present, they are highly decentralized in the sense that they respond only to local error measurement. The slower controls evolve at the rate kT_s and are in support of frequency and voltage regulation. As such, they are centralized at each subsystem level. Their purpose is also to stabilize the tie-line flows to the values scheduled from the higher system level at the rate KT_t. The slowest is the tie-line flow regulation for expected load variations over the time horizons KT_t. At present, the latter is not an automated systems control function.

These technical activities create sequences of control actions at each subsystem (control area) level

$$P_{Gi}^{secondary}[kT_s], P_{Gi}^{secondary}[2kT_s], \quad \cdots \quad i \in N_A^s \quad (3.5)$$

and at the interconnected system level

$$P_{Gi}^{tertiary}[KT_t], P_{Gi}^{tertiary}[2KT_t], \quad \cdots \quad i \in N_A^t \quad (3.6)$$

Similar sequences of control actions are generated in real time for reactive power/voltage regulation at each subsystem level

$$Q_{Gi}^{secondary}[kT_s], Q_{Gi}^{secondary}[2kT_s], \quad \cdots \quad i \in N_A^s \quad (3.7)$$

---

[3]Observe analogous time frame decomposition in modeling the members of the N_C set and the N_A set. This is necessary, since for each N_C change it is required to have control that balances the system over the time frame of interest according to the performance criteria agreed on.

[4]Voltage control is only automated in Europe [46].

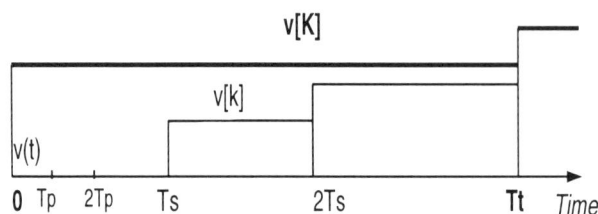

Figure 3.4: Control-induced time scales for operations planning

and at the interconnected system level

$$Q_{Gi}^{tertiary}[KT_t], Q_{Gi}^{tertiary}[2KT_t], \quad \cdots \quad i \in N_A^t \tag{3.8}$$

Here N_A^p, N_A^s, and N_A^t indicates the set of generators participating in real power/frequency and the reactive power/voltage regulation over various time frames. It is relevant to observe that the secondary-level control allows for autonomous generation scheduling at each control area level as long as the tie-line flows are maintained accordingly. Control-induced time scale separation for current load following is shown in Figure 3.4.

We suggest that these systems control services should meet certain well-defined performance criteria independent of the competitive nature of supply/demand in the set N_C not participating in these functions. These performance criteria are described next.

3.4 Performance criteria at the subsystem level

For completeness and a clear distinction between the roles of different performance criteria, we first review here performance criteria well known in the power engineering literature.[5]

Many of these are routinely used at present by the utilities in their control centers for resource scheduling for expected loads, and for on-line regulation in response to load uncertainties and other uncertainties on the system.

The material in this chapter differentiates between the objectives of meeting known nominal demand in a static manner, on the one hand, and the objectives of dynamically regulating system outputs, such as frequency and voltage, in response to unknown deviations, on the other.

---

[5] A degenerate case of a subsystem level within an interconnected system is an isolated power system not connected to any other subsystems. The analysis used for the subsystem level is therefore also useful for this case.

3.4.1 Static optimization objectives

The performance criteria for unconstrained economic dispatch, constrained real and reactive power economic dispatch, and optimal power flow are described first (Section 3.5). All these methods are static optimization tools, which are derived under the assumption of no dynamic operating problems. As such, they are not ready to be used for dynamic regulation specifications. Two inherent features of existing static optimization tools should be kept in mind: (1) They are suboptimal relative to the performance achievable when system regulation is done over longer time horizons for expected demand; and (2) they usually require full knowledge of the system state, in contrast to many dynamic regulation schemes, which rely on minimal information exchange relevant only to the process of interest.

The most significant characteristic of static optimization objectives is that they are not intended for dealing with the dynamic performance of the system. They use single snapshot data without correlating them to the system performance prior to this computation. Thus, static tools are not used for controls needed to ensure certain dynamic performance over mid- or long-term horizons.[6]

Up to this time, power systems behaved more or less as static systems, for which the system inputs (load variations, in particular) could be predicted quite accurately. The use of static optimization tools for scheduling generation for the anticipated load was sufficient in this case.

Significant contingencies on the system are dealt with in a preventive operating mode, by which scheduling is done to remain within the specifications if any single event violates the $(n-1)$ security criterion. This results in more conservative scheduling than without the $(n-1)$ security criterion constraint. In a competitive environment the use of the $(n-1)$ security criterion will definitely have to be reconsidered, to prepare the system for dealing with drastic equipment emergencies. However, even in the normal operating mode in a competitive environment, because of very significant load/generation uncertainties, enhanced on-line optimization controls in real time will be required.

3.4.2 Dynamic optimization objectives

Performance criteria formulated for the purpose of maintaining frequency and voltage quality over mid- and long-term horizons despite significant system input uncertainties are described next (Sections 3.7 and 3.8). The performance criteria for characterizing system response in a dynamic manner offer

---

[6]We are not concerned here with transient, extremely short-term stabilization via local controllers distributed throughout the system. The concern is with resetting their set values so that the performance at both subsystem and interconnected system levels over longer-term horizons is as specified.

qualitatively different measures than the static performance criteria. For example, the sum of the values of performance criteria sampled at each time instant kT_s gives a dynamic measure of system performance quality. This is not the case with static measures.

In the United States the only actively practiced on-line control scheme of this type is for systemwide frequency regulation, known as the load frequency control (LFC)/automatic generation control (AGC) scheme. The performance criteria used for LFC/AGC are important to understand for the future enhancement and valuing of automated control of interconnected power systems.

Dynamic generation-based voltage regulation, i.e. scheduling of reactive power/voltage resources to meet load variations over mid- and long-term horizons according to prespecified performance criteria, is not practiced at present in the United States. Voltage regulation is done only in an asynchronous manner using static optimization tools. In Europe, however, both the French and Italian electric power systems have automatic secondary voltage control (AVC) schemes for regulating reactive power out of generators. These schemes are designed according to performance criteria that guarantee prespecified quality of voltage response over mid- and long-term horizons.

We review these performance criteria as examples of the criteria used for optimal voltage control design. For the U.S. power system to operate in a satisfactory manner under competition, it may become relevant to require meeting performance objectives for dynamic voltage regulation over mid- and long-term horizons. Otherwise, no incentives will exist for these technical functions.

A major emphasis in this text is on the need to maintain electric power performance over mid- and long-term horizons according to well-defined performance criteria. None of the static optimization tools can provide systemwide frequency specifications over long time horizons, i.e., they are not applicable to dynamic regulation problems such as AGC and AVC.

The dynamic nature of models for measuring system performance is fundamental for relating technical changes on the system to performance-based economic value. Most of the value of transmission services and the supporting control actions are cumulative in nature over mid- and long-term horizons. Dynamic economic models are needed for studying performance-based cost and price feedback design according to optimal control designs for dynamic systems. Stability properties of costing mechanisms and the existence of equilibria of the economic processes can only be studied by means of dynamic models. It is possible to define optimal technical processes (and therefore make statements about the best technical activities on the system) only if the dynamic process is well defined.

3.5 Static optimization in an open access system

This section briefly reviews the present optimization objectives in operating large electric power systems and discusses the changes in these objectives that would be necessary under deregulation.

At present, generation is scheduled at a systemwide level with the single objective of minimizing total cost for a given demand. This is referred to as the *ideal technical efficiency* of generation production. To relate this performance criterion to the conventional notions of economic efficiency, the definition and role of social welfare maximization are briefly reviewed in the context of electric power industry operation.

In a centralized industry structure, where cost/benefit functions are fully known, it is straightforward to generalize existing optimal dispatch algorithms, which were formulated to achieve technical efficiency on the generation side, to incorporate benefit maximization and achieve ideal economic efficiency. Benefit is just viewed as the negative cost of power on the demand side [38].

In a competitive environment, the emphasis is on partially distributed decision making and on performance objectives of individual market participants whose cost/benefit (utility) functions are generally not known at a centralized level. For such an environment, distributed optimization objectives are defined first. The achievable systemwide efficiency under open access relative to ideal economic efficiency is analyzed. This analysis leads to a conclusion to retain coordination of technical management and cooperative pricing for ancillary generation in support of a competitive supply/demand market. Systems control services are defined as those services in a distributed power system that work to keep the system together (operating safely and efficiently) for the benefit of all users [21].

Section 3.5.1 describes the assumptions under which present centralized optimal dispatch is conducted, and Section 3.5.2 gives the formulae for generation cost minimization. In Section 3.5.3 we review the basic costs of keeping the system together at present. This cost is well known to engineers but not necessarily common knowledge. The material is written in the context of preparing the ground for further understanding of the principles of nondiscriminatory and efficient pricing methods under competition. As part of this overview, a modification of the optimal dispatch problem is introduced that does not require sudden jumps in prices when operating constraints are reached. This simple change can be shown to be basic to cost allocation of ancillary generation to the generation side of the utility business instead of to transmission owners, as suggested in some recent literature.

Section 3.5.4 briefly describes achievable technical efficiency in the regulated industry. Sections 3.5.5–3.5.7 point out assumptions that do not hold for a deregulated industry and define basic optimization objectives relevant for a competitive electric power industry. Formulae for performance criteria of individual market players are provided. In the same sections, a brief review of conditions for ideal economic efficiency in terms of social welfare maximization is given. It is shown that a straightforward generalization of existing optimal dispatch can incorporate customers into an (economically) efficient centralized energy management scheme. Metering and communications requirements, on the other hand, may be quite costly and should be carefully assessed.

In Sections 3.5.8–3.5.13 the integration of distributed objectives of market participants with systemwide ancillary generation is pictured as a hierarchical system driven by the competitive market and coordinated for efficiency and nondiscriminatory pricing by signals from the ancillary services. Section 3.6 discusses static optimization for horizontally structured systems.

3.5.1 Some assumptions under which present optimal scheduling algorithms are designed at the system level

The electric power industry is equipped at present with various optimization tools that are routinely used for generation scheduling needed to meet anticipated demand. Tools such as unconstrained economic dispatch (ED), constrained economic dispatch (CED) [42], and optimal power flow (OPF) algorithms [43] are used for this purpose.

It is important to recognize several underlying assumptions that make static optimal dispatch effective:

- **Assumption 1:** Generation management is fully centralized at the Energy Management Systems (EMS) level.

- **Assumption 2:** Demand is known with high certainty. Only insignificant deviations from the known demand profile are expected, and they are compensated for in an automated way by closed-loop control schemes such as AGC.[7]

- **Assumption 3:** No strong distinction is made between ancillary generation and the rest of the generation.

- **Assumption 4:** Static optimization tools are sufficient for systemwide efficiency, assuming demand is known with high certainty (assumption 2).

---

[7]Typical data are shown in Figure 3.1.

3.5.2 Generation cost minimization: Ideal technical efficiency

The optimization problem here is the problem of minimizing the total generation cost

$$\sum_{i=1}^{N_G} c_i(P_{Gi}), \quad i \in N_G \tag{3.9}$$

for given generation

$$P_{Gi}^{min} \leq P_{Gi} \leq P_{Gi}^{max}, \quad i \in N_G \tag{3.10}$$

such that

$$\sum_{i=1}^{N_G} P_{Gi} = \sum_{j=1}^{N_D} P_{Dj}, \quad j \in N_D \tag{3.11}$$

The basic version of unconstrained economic dispatch finds a solution to this optimization problem for a system of arbitrary size. A necessary condition for solving this problem is known to be [41]

$$\frac{\partial c_1}{\partial P_{G_1}} = \frac{\partial c_2}{\partial P_{G_2}} = \cdots = \frac{\partial c_{N_G}}{\partial P_{G_{N_G}}} = \lambda^* \tag{3.12}$$

This condition is considered to define ideal technical efficiency in operating electric power systems. The term λ^* is known as the short-run marginal cost (SRMC) and at the optimum of (3.9) is the same for all generating units in this case. This ideally achievable cost of specific generating units is denoted by c_i^*.[8]

3.5.3 Basic operating cost of keeping the system together

At present great care is taken in the planning stages of both generation and transmission to operate close to ideal technical efficiency, i.e., that most of the units operate at the same SRMC λ^* as defined in (3.12) for the anticipated demand. This is a meaningful task, assuming that assumptions 1 and 2 are closely met in the actual system operation.[9]

The prime role of a transmission grid is to serve as a path provider for power flows between generation and demand buses, and to provide voltage/reactive power support by means of distributed devices such as switching

---

[8] Subscript i is carried here to emphasize the dependence of the optimal cost on the generation location.

[9] It is interesting to observe that the major stranded cost facing utilities at present is a consequence of drastic demand deviations from the anticipated. This change is primarily the result of external forces outside of the utilities' control.

capacitors sprinkled throughout the system. Neither of these two functions has significant operating cost. Once the equipment is in place, the cost of operating is relatively small.[10]

In contrast to this, a high operating cost can be experienced on the system for so-called ancillary generation. To avoid confusion, it is important to recognize that the term *ancillary services* as recently used in some public documents [21], has a somewhat different meaning from that typically understood in existing operation. This difference is elaborated on throughout this text.

At present economic dispatch/scheduling, loss compensation, and load following are integral services provided by all generating units participating in economic dispatch for the anticipated demand. Only deviations from anticipated demand caused by either small random fluctuations or significant outages rely on ancillary generation. Typically a large system has only a handful of units directly dedicated to systemwide regulation in response to relatively small random variations. For frequency regulation, these units are known as the AGC units. The generation reserve for system protection in the event of major outages, on the other hand, is planned in such a manner that the most economical units are used whenever possible in the actual operation.

Loss compensation and its cost

It has been fully documented in the engineering literature that when real power transmission loss is compensated for while attempting to solve an unconstrained economic dispatch as defined above, the optimality condition (3.12) becomes a condition

$$\frac{\frac{\partial c_1}{\partial P_{G1}}}{1 - \frac{\partial P_{loss}}{\partial P_{G1}}} = \frac{\frac{\partial c_2}{\partial P_{G2}}}{1 - \frac{\partial P_{loss}}{\partial P_{G2}}} = \cdots = \frac{\frac{\partial c_{NG}}{\partial P_{G,NG}}}{1 - \frac{\partial P_{loss}}{\partial P_{G,NG}}} \quad (3.13)$$

The marginal cost of each unit after compensating for systemwide transmission loss P_{loss} is denoted here by $\lambda_i^{loss} = \frac{\lambda^*}{1 - \frac{\partial P_{loss}}{\partial P_{Gi}}}$, and it generally depends on the generation location i.

The operating cost of each unit i is obtained by multiplying the coefficient λ_i^{loss} and the amount of P_{Gi} at the optimum. For purposes of further discussion we denote this cost by c_i^{loss}. This leads to a formal definition:

- **Definition 3.1:** *Cost of loss compensation Δc_i^{loss} is the difference between the cost achievable when loss is compensated for c_i^{loss} and the ideally achievable cost c_i^*, given all other conditions identical.*

---

[10] Only operating cost is of interest in this section.

Chapter 3   37

The actual values of Δc_i^{loss} are non-uniform at different buses.

For later discussion it is important to recognize that this cost component is a function of *all* system states (voltages and currents), and that as such it is not computable at the individual buses. In other words, the transmission loss cost as seen at each bus generally depends on both demand and generation at all system locations.

Ancillary generation-based cost of reactive power/voltage support

Without getting into the specific mathematics, suffice it to say that when economic dispatch as defined above is required to provide for voltages at all buses throughout the system to remain within the operating constraints, i.e.,

$$V_i^{min} \leq V_i \leq V_i^{max}, \quad i \in N_G, N_D \tag{3.14}$$

an additional cost is borne. A typical numerical tool for computing such cost is OPF [41]. Standard OPF formulation typically treats constraint (3.14) as a hard constraint by defining a Lagrangian of the form

$$\begin{aligned}\mathcal{L}_{volt}(\underline{P_G}, \underline{V_G}) &= \Sigma_{i \in N_G, N_D} c_i(P_{Gi}) + \\ &\quad \Sigma_{i \in N_G, N_D} m_i(V_i - V_i^{min}) + \\ &\quad \Sigma_{j \in N_G, N_D} n_j(V_j^{max} - V_j)\end{aligned} \tag{3.15}$$

The optimization is done with respect to controls $\underline{u}$, which are in this case $P_{Gi}, V_{Gi}, i \in N_G$. Voltages at loads and phase angles throughout the system (except at the slack) are system variables $\underline{x}$ subject to the load flow constraints. For purposes of further discussion, the load flow equations are represented symbolically as

$$\underline{g}(\underline{x}, \underline{u}) = \underline{0} \tag{3.16}$$

Using this problem formulation, functions $m_i, n_i, i, j \in N_G, N_D$, have a well-known interpretation in terms of out-of-merit real power deviations from optimum cost λ^* for voltage regulation needed [43, 102]. They typically are taken to be zero as long as the limits are not violated, i.e., they are treated as hard constraints.

A close look into the optimization techniques suggested in the past [45, 46] reveals that the constraint (3.14) need not be formulated as a hard constraint. Instead, the optimization problem for minimizing generation cost while observing voltage constraints can be formulated as an em inequality-unconstrained control problem, with the performance objective being

$$\begin{aligned}\mathcal{L}_{volt,soft} = &\ \Sigma_{i \in N_G} P_{Gi} + \\ &\ (\underline{V} - \underline{V}^{min})^T Q_{min}(\underline{V} - \underline{V}^{min}) + \\ &\ (\underline{V}^{max} - \underline{V})^T Q_{max}(\underline{V}^{max} - \underline{V})\end{aligned} \qquad (3.17)$$

subject only to equality constraints (3.16). Matrices Q_{min} and Q_{max} are weighting matrices and can be chosen to accommodate relative voltage quality throughout the system [46]. $\underline{V}$ is a vector representing voltages at both generator and demand buses. Problem formulation (3.17) can be viewed as allowing for soft voltage constraints.[11]

While it may not be clear at this point why two seemingly similar problem formulations are described, this will become apparent in the context of dealing with transmission line constraints.

The cost seen at each generating unit after solving this optimization problem is denoted here by c_i^{volt}.[12] The cost difference $\Delta c_i^{volt} = c_i^{volt} - c_i^*$ is the cost seen by each generating unit necessary to regulate both generation and demand voltages throughout the system.

It can be shown that when only generator voltages are regulated, component Δc_i^{volt} can be computed locally at each generator bus. However, when both generator and load voltages (being part of the states $\underline{x}$) are regulated, the cost function (3.17) becomes a *qualitatively new* problem relative to the problem of solving unconstrained dispatch given as (3.9). This qualitative difference results from the fact that the general form of the performance criterion (3.9) is

$$minimize \ c(\underline{u}) \quad wr. \ to \quad \underline{u} \qquad (3.18)$$

Performance objective (3.17) is, on the other hand, a problem of the form

$$minimize \ c(\underline{u},\underline{x}) \quad wr. \ to \quad \underline{u} \qquad (3.19)$$

and subject to (3.16).

It is documented throughout this text that this is the major qualitative difference between the tasks of ancillary services, on the one hand (problem (3.19) subject to (3.16)), and straight cost/benefit optimization (problem (3.18)), on the other.

While the systemwide performance objective (3.9) does not reflect explicitly the cost of reactive power loss compensation, significant cost deviations in delivering real power could occur for large deviations in voltage when optimizing either (3.15) or (3.17) [46]. A performance criterion of the type (3.15)

---

[11] This may be a good feature for avoiding sudden jumps in electricity prices.
[12] No closed-form solution to the constrained OPF exists.

is not the only type of static performance criteria used for changing settings of generator voltages. The choice of best performance criterion for voltage has been a topic of much research in the past [60, 61]. The problem is acute in that even the basic notion of voltage optimality has not been standardized [62]. The conclusions are system-dependent [63], and are not readily generalizable. The choice of performance criterion for generation-based reactive power/voltage support affects the optimality of total generation cost when the limits on generation are exceeded. No explicit, agreed upon, cost curve for generation-based reactive power support as a function of reactive power is known at present. Moreover, some believe that because the cost of reactive power support implementable on the grid (instead of produced by generators) is relatively low, the control of voltage within the allowable limits could be done rather inexpensively [64]. This problem is fairly complex and more work is needed prior to standardizing the value of reactive power support on an interconnected system. If this is not done, otherwise most attractive economic transactions might not be feasible. For further understanding of the unique features of reactive power propagation on an interconnected system the interested reader could read [65, 66].

Ancillary generation-based cost of eliminating transmission constraints

Somewhat costly deviations from optimal dispatch are seen when some of the transmission constraints are violated. In some recent literature this cost is referred to as a congestion cost [27, 38].

In this case, deviations from the ideally achievable cost c_i^* are necessary in two distinct situations:

1. The transmission grid is not capable of distributing injected generation required for particular demand specifications.[13]

2. Operating constraints on system outputs, such as the allowable line flows, are exceeded. These constraints are expressed as

$$P_{ij} \leq P_{ij}^{max}, \quad i,j \in N_G, N_D \tag{3.20}$$

Typical problem formulation is based on treating constraints (3.20) as hard constraints and defining a Lagrangian of the form

$$\begin{aligned}\mathcal{L}_{cong} &= \Sigma_{i \in N_G} c_{Gi}(P_{Gi}) + \\ &\quad \Sigma_{i \in (N_G, N_D)} \Sigma_{j \in (N_G, N_D), i \neq j} \mu_{ij}^{max}(P_{ij} - P_{ij}^{max})\end{aligned} \tag{3.21}$$

---

[13]It is known that any electric network cannot transfer more power than specified by the conditions of the maximum power transfer theorem. In an electric power network this problem is well known as the problem of not having a solution to the load flow problem [47]. The cost of this is not discussed in this section, since this problem may be viewed as an emergency operation issue.

The (nonclosed form) solution to this problem provides total cost at each bus i c_i^{cong}. The difference $\Delta c_i^{cong} = c_i^* - c_i^{cong}$ is then the exact out-of-merit generation cost required to keep power flows within the transmission limits.

There has been a great deal of controversy over using coefficients μ_{ij}^{max} for pricing under competition [38]. To avoid potential misunderstanding when using μ_{ij}^{max} as indicators of congestion costs [38] we propose a reformulation of the optimization problem of interest. Similarly to the way of voltage constraints are dealt with as either hard constraints (formulation (3.15)) or soft constraints (formulation (3.17)), we propose to pose the problem of generation cost minimization as an *inequality-unconstrained* problem of the form

$$\begin{aligned}\mathcal{L}_{cong,soft} = & \ \Sigma_{i \in N_G} c_G i(P_{Gi}) + \\ & (\underline{P_{ij}}^{max} - \underline{P_{ij}})^T R_{max} (\underline{P_{ij}}^{max} - \underline{P_{ij}})\end{aligned} \quad (3.22)$$

The optimization is done with respect to $P_{Gi}, i \in N_G$, and subject only to the load flow equality constraints. Weighting matrix R_{max} is a diagonal matrix whose terms reflect relative normalized susceptances of individual transmission lines. $\underline{P_{ij}}$ is the vector of all real power line flows through transmission lines connecting buses ij. In the actual numerical method developed to optimize (3.22) these terms can be made to be *operating-conditions-dependent*: as the line flows approach their limits, the coefficients increase exponentially. This is a well-understood method of penalty function used for dealing with constraints [48].

A variety of well-tested algorithms for optimal dispatch that impose equality load flow constraints and inequality constraints on system output variables are used by utilities at present to ensure optimum dispatch that is feasible and within operating constraints (3.20) [49]. A relatively minor modification of the present optimal dispatch is needed to convert the problem into the soft optimization defined as in (3.22).

Note: As in the case of the other system-related costs Δc_i^{loss} and Δc_i^{volt}, Δc_i^{cong} reflects the entire situation on the system. It is therefore *not* meaningful from an engineering point of view to define a distributed notion of a path-by-path congestion cost within the given grid [38, 50]. While this would be somewhat desirable in a deregulated environment, it is essential to recognize that the "contract path" of any type cannot be based on solid engineering notions, and as such it directly leads to the problem of potential discrimination for profit.[14]

This leads us to a basic definition:

---

[14] It is essential to understand that the congestion cost is not an indicator of any operating costs borne by the grid. This cost is truly generation-based and is reflected in the Δc_i^{cong} quantity. If this is not kept in mind, many accounting problems may arise in a so-called vertically disintegrated environment.

- **Definition 3.2:** *The out-of-merit cost* at each generating unit for keeping the system together in response to *scheduled* system demand can be expressed as[15]

$$c_i^{together} = \Delta c_i^{loss} + \Delta c_i^{volt} + \Delta c_i^{cong} \tag{3.23}$$

This is the total cost of out-of-merit generation dispatch necessary to compensate for transmission loss, meet static security constraints, and make the transaction physically possible in addition to the cost component c_i^* of meeting the given demand at lowest cost defined in (3.9). Depending on the actual system conditions, this cost ranges from zero to significant. Specific numerical techniques that compute feasible solutions for a large system under present operating assumptions are readily available and are used by major utilities throughout the world [49].

In sum, computing specific cost components needed to keep the system together is straightforward by means of present optimal dispatch. Furthermore, since under assumption 2 the demand side does not change in response to generation cost, optimizing cost on the generation side is equivalent to achieving overall efficiency, including for the customers.

3.5.4 Achievable technical efficiency in the regulated industry

It is easily demonstrated that if demand is time-varying but certain, these static optimization tools can achieve near-optimal generation use without sacrificing much efficiency over long time horizons. In other words, SRMC (result of static optimization) almost recovers long-run incremental cost (LRIC) [28].

Claim 1: The least-cost total generation defined in (3.9) is achieved in an optimal manner when the cost of keeping the system together $c_i^{together}$ (3.23) is optimized using present centralized optimal dispatch.

A rigorous proof for this is based on the result that a general performance function $c(\underline{x}, \underline{u})$ optimized with respect to control vector $\underline{u}$ subject to equality constraints $\underline{g}(\underline{x}, \underline{u}) = \underline{0}$ is highest when optimization is done in a centralized manner, i.e., when the entire vector $\underline{u}$ is optimized simultaneously instead of in some distributed manner. This is a general result underlying many of the pros and cons for specific industry structures, and it should be carefully kept in mind.

Claim 1 is made to emphasize that present centralized (fully coordinated) dispatch is as efficient with respect to performance criterion (3.9) as one can hope for. This is true even when the cost of keeping the system together is added. In other words, one cannot make the argument that if transmission

---

[15] Formula (3.23) should not be interpreted to imply additivity of its particular cost subcomponents at the optimum. It is used to indicate cost components associated with keeping the system together as a function of constraints observed.

loss were compensated for locally, the total cost $(c_1^{loss} + \ldots + c_{NG}^{loss})$ defined in (3.13) could be further reduced. Similar arguments hold for present technical efficiency in managing operating constraints such as (3.14) and (3.20). Going back to the world of fully distributed homeostatic controls as a means of keeping the system together, for example [51], it is straightforward to show that *any* implementation of this sort would reduce systemwide technical efficiency. This is not to say that it would not, potentially, reduce some specific $c_i^{together}$ relative to the cost computed by centralized optimal dispatch. This rather subtle distinction may become relevant in the future.

It is worthwhile observing that the arguments concerning systemwide efficiency when applied to a multi-utility environment speak strongly in favor of more coordination over large geographic areas by means such as regional transmission groups [21].

3.5.5 Assumptions that do not hold in a deregulated industry

At present, most large electric power systems are characterized by net generation excess. This results in competition for customers on the generation side and competition for the most beneficial supplier on the demand side.

In this situation, generation that does not take part in systemwide regulation is seen as price-driven, and does not lend itself to conventional scheduling of centralized economic dispatch at a system level. This directly violates assumption 1 which underlies existing optimal dispatch algorithms (see Figure 3.2).

The variations in native load will be more dynamic and harder to predict than in the past, particularly if retail wheeling occurs. This violates assumption 2, that demand be known with high certainty.

Ancillary generation designated for systemwide regulation may not be owned by the system. If, for example, independent power producers (IPPs) want to participate in systemwide regulation, they will have to be incorporated into the existing control architectures in a nondiscriminatory manner. This requires a sharper distinction between a generating unit participating in systemwide regulation and a strictly profit-driven generation unit. Therefore, assumption 3 does not hold. Thus, assumption 4, that static optimization tools are sufficient, is no longer self-evident.

In summary, the assumptions under which static optimal dispatch is now implemented cannot be counted upon in a competitive environment. This calls for evolution of present optimal dispatch into a tool capable of accommodating distributed industry with the highest possible efficiency and in a nondiscriminatory manner. The remainder of this text investigates possible evolution of this type.

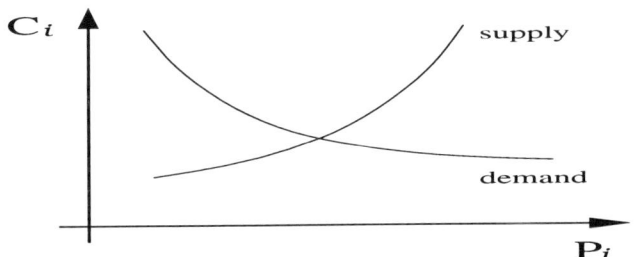

Figure 3.5: Typical supply and demand curves

3.5.6 Need for relaxing the demand-related assumptions (2 and 4)

Because of demand becoming more responsive to energy prices, it is essential to adjust present optimization tools at the EMS level to accommodate utility "nonmembers" (price-competitive generation and demand).[16] At the same time, it remains essential to do this in an efficient manner. Typical static curves representing these two functions are shown in Figure 3.5.

Any (generation) cost-based price signal is fundamentally deficient in giving the right incentives for most optimal demand.

Social welfare maximization using present optimal dispatch: Ideal economic efficiency

It is straightforward to allow for total benefit maximization on the demand side at the same time as total generation cost is minimized [38].

This problem is basically formulated as the problem of maximizing social welfare benefits:

$$Social\ welfare = -(\Sigma_{i \in N_G} c_{Gi}(P_{Gi}) + \Sigma_{j \in N_D} c_{Dj}(P_{Dj})) \quad (3.24)$$

for given generation limits

$$P_{Gi}^{min} \leq P_{Gi} \leq P_{Gi}^{max}, \quad i \in N_G \quad (3.25)$$

and demand variation limits

$$P_{D_i}^{min} \leq P_{D_i} \leq P_{D_i}^{max}, \quad j \in N_D \quad (3.26)$$

such that

$$\Sigma_{i=1}^{N_G} P_{Gi} = \Sigma_{j=1}^{N_D} P_{Dj} \quad (3.27)$$

---

[16]One should recognize that in concept there is no qualitative difference between price-responsive generation and demand; they only differ in the sign of power used.

Theoretically, this formulation is a direct generalization of unconstrained economic dispatch.

The optimum is defined by conditions very similar to the conditions in (3.12), except that benefit optimization is also achieved. As expected, it is easily shown that both generation and demand operate at the same SRMC:

$$\frac{\partial c_{G1}}{\partial P_{G1}} = \cdots = \frac{\partial c_{NG}}{\partial P_{NG}} = \cdots = \frac{\partial c_{D1}}{\partial P_{D1}} = \cdots = \frac{\partial c_{ND}}{\partial P_{ND}} = \lambda_{sw}^* \qquad (3.28)$$

Term c_{Dj} stands for negative benefit on the demand side at bus j. It is important to observe that this generalization is useful only to the demand interested in responding to cost signals. The solution to this optimization problem is referred to here as *ideal economic efficiency*. A direct generalization of present tools to optimize total social welfare in a centralized manner can be made, assuming the demand benefit functions are known at the EMS level in addition to the generation cost functions; one optimizes a performance criterion with respect to both cost-responsive generation and demand.[17]

3.5.7 Need to reconsider the performance objectives (assumptions 1–3)

The distributed energy market can be viewed as a noncooperative game for profit by individual players. Several general facts should be kept in mind:

- Cost/benefit functions of individual participants are not known to the others.

- In a nonperfect market, when information on power quantity and cost functions is not known at all or only partially known, there is a significant difference between cost and price. Only at an economic equilibrium do these two quantities become equivalent [52].

- Power trade in a market generally takes place whenever quantities (power) balance but not necessarily at the economically most efficient operating point.

It therefore becomes necessary to differentiate between performance objectives for economic efficiency and for profit optimization.

To start with, one recognizes that a performance objective of an individual market participant i (generation or demand) can be expressed as the problem of maximizing profit:

$$\pi_i(P_i) = p_i(P_i)P_i - c_i(P_i) \qquad (3.29)$$

---

[17]The cost of actual implementation of these conditions is quite a different question, since at present no hardware has been implemented for active demand regulation at the EMS level.

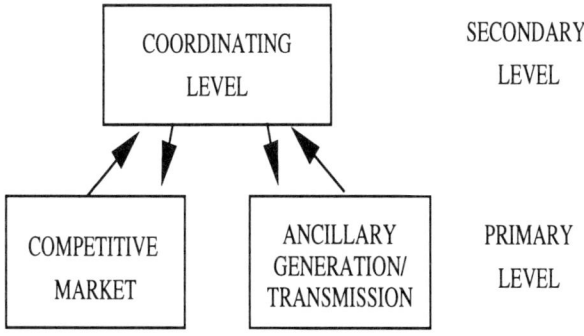

Figure 3.6: Single utility viewed as a two-level hierarchy

Here π_i stands for the profit of a market participant i, and p_i stands for the price obtained through some sort of (centralized or bilateral) bidding process. c_i is the actual cost of producing generation P_i. All three variables are nonlinear functions of power quantity P_i. The basic objective at each market participant level is to maximize profit by

- Maximizing the revenue $p_i(P_i)P_i$ of quantity P_i sold through competitive bidding to someone in the market.

- Minimizing cost $c_i(P_i)$ of producing power P_i.

Performance objective (3.29) does not recognize any cost $c_i^{together}$ borne by the need to facilitate the market transactions, unless specific rules are imposed. For nondiscriminatory pricing this cost must be accounted for.

3.5.8 Hierarchical structures in a distributed industry

In attempting to formulate optimal methods for keeping the system together in response to competitive supply/demand (i.e., when assumptions 1–4 are no longer valid), we first observe a conceptually new information structure.

Two types of information structure are of direct interest; one set is relevant for technical operation under competition, and the other is for pricing. They cannot be assumed to be the same because we are dealing with the process of gaming for profit on the competitive generation side, demand-side management for benefit maximization, and ancillary generation trying to define itself in this new process. The objectives of competitive supply/demand participants are *highly distributed*, with ancillary generation trailing far behind in defining its value to the market.

For purposes of further analysis in this section, we view this organization as a hierarchical system whose changes are driven by the market. In a single-utility setup under open access, one has at least a two-level hierarchy. As shown in Figure 3.6, at the primary (lower) level competitive supply/demand enters the system as one type of market. The second market at the same level represents ancillary generation, whose basic function is to coordinate systemwide performance. The role of the secondary level is strictly coordination, and it can be interpreted as an independent system operator (ISO), whose functions are not fully defined at this point. The need for its existence and the functional details are studied in the remainder of this text. It is helpful to observe that a particular case of this two-level structure is the present (secondary-level) system coordination of the *single* primary level comprising the entire demand and generation.

3.5.9 Achievable efficiency under open access–ISO market level

This problem is divided into two subproblems to relate directly to the two-level hierarchy shown in Figure 3.6. The first portion of the primary level is concerned with competitive supply/demand. These are defined as system inputs in Section 3.1. The second is the ancillary market, which is needed to keep the system within constraints under open access by means of various generation-based controls. The efficiency of the ancillary market is strongly dependent on the principles of systems control for responding to various market uncertainties. Systemwide efficiency is analyzed by attempting to understand what is the most efficient achievable under competition.

3.5.10 Achievable efficiency of competitive supply and demand

The first question to ask is, What does the process of individual profit making have in common with the systemwide social welfare optimization criterion (3.24)? A direct answer can be found in general economics under certain assumptions, such as the absence of externalities between the economic agents; perfect information of consumers about the product; and the private nature of the goods. Under these conditions the following holds:

Claim 2: At an economic equilibrium every player operates at the point that is also a systemwide optimum.

The proof for this claim is widely known and so is omitted here. It is a direct consequence of thinking that one increases the quantity traded only as long as nonzero profit is made [52].

It is important to keep in mind, however, that a potential problem exists in applying this result to the present power markets because the system is not nearly at its ideal economic equilibrium. Instead, much room exists for profit making under the present rules, particularly by informed market players. Effective profit making in nonefficient (nonperfect) markets requires knowledge about the utility functions of other players, on both the supply and demand sides. These functions (cost and benefit functions c_{Gi} and b_{Dj}, respectively) will not be made available unless explicitly required by the regulations.

Nevertheless, this is an important result that is often used as the basic argument for justifying systemwide efficiency under competition.

For our purposes, we use this result to reiterate that in a perfect competitive supply/demand market similar systemwide efficiency can be obtained as in the present centralized system. This is, furthermore, an argument that can be used by advocates of both bilateral and coordinated pricing to claim that for unconstrained system operation there is no genuine efficiency gain from competition or from coordination. This argument seems to be forgotten as we witness endless discussions on this topic [53]. (Given an imperfect market, there is an argument for coordinated pricing of systems support services over bilateral.[18])

3.5.11 Achievable economic efficiency of generation-based systems control

We observe next that the social welfare criterion in its basic form does not recognize either of the two unique principles of power system operation, i.e., its spatial or its temporal aspect. This leads to possible inefficiencies as seen from a systemwide (secondary) level when using the social welfare criterion blindly. The spatial aspect of energy cost and its economic value require a non-uniform price at different electrical (and geographical) locations in the system. The marginal cost of energy supply and consumption depends on its location in a large system and also on the type of supply. It is therefore intuitively clear that much of the achievable efficiency in deregulated power systems will be determined by how efficient ancillary services are. This leads us to the main result in the following section.

3.5.12 Need for coordinated generation-based systems control in support of competitive markets

It is straightforward to understand that the competitive supply/demand market at the primary level can be viewed as the trigger of technical activities of generation-based systems control at the same level. The problem than becomes one of optimizing the cost of generation-based systems control of the form (3.9) and ensuring that systemwide operating constraints are met.

---

[18]The question is more complex than this; see [52].

In this sense, ancillary generation takes on the role of keeping the system together in response to profit-driven changes in the supply/demand market according to prespecified performance objectives at various levels of the nested hierarchy. These costs are computed according to the same formulae as in the case of centralized industry structure, according to the values $c_i^{together}$ defined in (3.23).[19]

As explained in Section 3.5.3, any performance criterion for the ancillary generation function must be viewed as a function of both controls $\underline{u}$ and system states on the rest of the system $\underline{x}$. This immediately brings the ancillary service problem into the category of optimization problems that are best solved at a systemwide level. Otherwise, the solutions are suboptimal relative to the ideally achievable cost minimization of systemwide ancillary generation.

Need for price coordination under open access

Here we omit detailed discussion of general results from theoretical economics. However, it appears that major recent work, such as [54]–[56], suggests that these services are best priced when they are viewed as being provided in a cooperative rather than a competitive mode.

To start with, solid grounds for making a service cooperative could be expressed by a simple inequality condition

$$\Sigma_{j=1}^{j=N_A} c_{Gj}(P_{Gj}) \geq c_A(P_A) \tag{3.30}$$

Here N_A is the number of generators participating in ancillary service to the competitive supply/demand market. $c_A(P_A)$ is the cost of coordinated ancillary service. Under this condition, the efficiency is optimized when a set of goods provides service in a cooperative, cost-based manner [55]. This is always true of ancillary service, given the basic nature of cost function (3.19).

3.5.13 Optimal structure for operating and pricing electric power systems under open access

Based on the preceding observations, the case could be made that for both technical and economic reasons the most efficient price signals are achieved when all contributors to ancillary generation are used in a coordinated manner for technical implementation and a cooperative manner for pricing. This is a direct justification for requiring a coordination level, as shown in Figure 3.6, for both technical and economic efficiency.

Based on the preceding analysis of the nature and cost of keeping the system together under competition, the following results are conjectured in this section:

---

[19]The real difference comes from a different charge allocation method than at present.

- **Result 1:** Systemwide optimization of technical (economic and engineering) efficiency achievable in a competitive energy market requires centralized scheduling and centralized pricing of ancillary service. This is a direct consequence of the result on optimality conditions for the general cost function (3.19) and treating ancillary generation as a common good to the competitive supply/demand market.

- **Result 2:** Even in an imperfect market centralized management and pricing of ancillary generation is likely to improve systemwide efficiency and allow for nondiscriminatory cost allocation. This is only a conjecture, not straightforward to prove.

3.6 Static optimization of a horizontally structured system

Consider an electrically interconnected power system consisting of administratively divided areas $1, 2, \ldots, R$, as shown in Figure 2.1. The interconnections among the areas, referred to as the tie lines, are typically designed to be electrically much weaker (and most often fewer in number) than the interconnections among generating units. Denote the generating units within area i by $G_1^i, G_2^i, \ldots, G_{n_g}^i$, and the loads within this same area by $L_1^i, L_2^i, \cdots, L_{n_l}^i$.

The reason the tie lines are electrically weaker that the intra-connections within each area is that individual power systems, as commonly operated, are designed to be autonomous. This means that each power system should be capable of meeting its own total area load reliably and not have to directly depend on a continuous energy exchange with the rest of the interconnected network of systems.

Energy exchanges with neighboring systems are scheduled only for economic reasons, or in situations when an area experiences an unplanned load increase or generation loss. The energy management of an electrically interconnected system in its normal operating mode is such that each area meets its own load while maintaining any scheduled energy exchanges with the neighboring areas across tie lines.

To meet these principles of energy management in a horizontally structured large electric power system one could, in principle, use identical static performance criteria as at a subsystem level. At present, in "tight" power pools, unconstrained economic dispatch is used for scheduling the resources on the entire system. More complex numerical tools, such as OPF, are not used for on-line operations of very large pools. Instead, it is common practice to perform optimization at the subsystem level, assuming that tie-line flows and voltages at the boundaries with other subsystems are maintained according to agreed-on schedules at each subsystem level.

Static optimization tools are very demanding with respect to data management. Because of this, it has been relatively hard to make them a routine standby tool in control centers. Interpreting data and implementing the results is often difficult. Many of these algorithms require scheduling actions everywhere instead of only at critical locations.

In light of this, use of static optimization for horizontally structured systems under open access must be handled with some caution. While, in principle, it is possible to treat the entire system as a unit and use OPF algorithms for scheduling generation in response to specified demand, it is critical to keep in mind the objectives specified at the system level versus the objectives specified at a subsystem level. These must be clearly distinguished for a competitive industry to have well-defined objectives and nondiscriminatory economic valuation.

For example, one can envision at least two solution scenarios for static optimization related to operating constraints, such as transmission line and voltage constraints:

- The responsibility is local at each subsystem level for eliminating operating constraints while allowing for, in effect, uncontrolled energy exchanges with other participants. This could result in significant variations of tie-line flows among subsystems. This mode resembles the way in which many systems operate at present, except that the energy exchanges are compensated for retroactively by each participant. This so-called inadvertent energy exchange [1, 4] at present is compensated for on a voluntary basis.

  We show later in this text that this scenario does not require coordinated regulation at the interconnected system level; decentralized regulation at each subsystem level is sufficient. It is necessary, however, to provide conditions under which such a decentralized regulation scheme will remain stable as seen at the interconnected system level. This is a theoretical challenge in its own right.

  This scenario does not require the existence of centralized futures markets, and it lends itself to supporting direct bilateral transactions among various levels within a nested hierarchy structure.

- The second scenario is for a control structure that would allow autonomy to each participant and at the same time regulate operating constraints at the interconnected system level for a specified energy exchange with the rest of the system. This prespecified energy exchange would take place within a scheme coordinated at the highest level of hierarchy.

  This scenario is better understood in the context of energy markets. Assume a futures market for energy is introduced at the interconnected, highest system level. The energy exchanges among the market participants are likely to be defined by the futures market on an hourly basis, for example. The main operating responsibility of system operators will be to ensure that the energy exchanges conform to market specifications.

While conceptually both control structures are possible, a closer analysis is necessary to understand the advantages and disadvantages of these two scenarios in terms of systemwide efficiency and reliability. This analysis is one of the main topics of this text.

The types of tools needed for meeting static and other kinds of performance objectives will depend on which of the two scenarios is used as the standard for a horizontally structured industry under competition.

3.7 Present criteria for mid- and long-term dynamic performance

In addition to the static scheduling based on economic dispatch, most utilities are equipped with an AGC scheme for automatic rescheduling of real power generation in response to slow load variations from the anticipated load for which the economic dispatch is computed. For purposes of further derivations in this text, let us restate the model representing slow load deviations from their expected values as[20]

$$P_{Li} = P_{Li}[kT_s] + P_{Li}[KT_t] \tag{3.31}$$

$$Q_{Li} = Q_{Li}[kT_s] + Q_{Li}[KT_t] \tag{3.32}$$

at each individual load

$$i = 1, 2, \ldots, N_L \tag{3.33}$$

Symbol P_L stands for the real power demand, and symbol Q_L stands for the reactive power demand. Load deviations are modeled as consisting of two different components, a faster component $P_{Li}[kT_s]$, to be managed at each utility level, and the slower load component $P_{Li}[KT_t]$, relevant for regulation at the interconnected system level. This load model is intended to support the present operating philosophy of scheduling at a subsystem level first, before asking neighboring subsystems for help. It is for this reason that typical sampling rate T_s for AGC is of the order of 2 seconds, and the tie-line flow rescheduling among different utilities is slower. At present, it is not automated, and it is done in an ad hoc asynchronous manner whenever necessary.

---

[20]It is assumed that the fastest load fluctuations are stabilized at a device level.

In light of Section 3.1, models (3.31) and (3.32) can be interpreted by viewing term $P_L[kT_s]$ as the noncompliance term of scheduled transactions and the random fluctuations in the native demand, and term $P_L[KT_t]$ as representing short-term scheduled transactions.[21]

At present, the requirements for dynamic performance over mid- and long-term horizons for an isolated electric power system are not specified explicitly. There exist local protection devices such as under-frequency relays for the protection of generators and large industrial loads, which disconnect them from the rest of the system if the frequency exceeds prespecified limits. Similarly, at a control center level there are no explicit specifications for systemwide dynamic performance over long-term horizons.

3.7.1 Criteria for load frequency control (LFC)/ automatic generation control (AGC)

Present performance criteria for mid- and long-term horizons are best described in the context of LFC/AGC performance objectives. Conventional LFC/AGC has the main objective to regulate frequency at each subsystem (control area) level to its nominal value as random, relatively small, fast load fluctuations $P_L[kT_s]$, $k = 0, 1, \ldots$ take place. At the same time, it has the objective to regulate the net real-power flow out of the subsystem to its scheduled value obtained by means of static optimization tools under the assumption that the demand is certain. Conventional LFC/AGC achieves these goals by responding to the deviations in an area control error (ACE) signal computed for each area i, $i = 1, \ldots, R$. This figure for the control error in an area has two distinct components. The first component accounts for deviations in the net tie-line flow interchange with all neighboring regions. In mathematical terms, if $\sum_k F_k^i(t)$ is the net sum of all tie-line flows into area i at time t, and $\sum_k F_k^{i^{sched}}(t)$ is the net sum of scheduled tie-line flows at that instant for the expected demand and generation in this area, then the first component of the ACE for area i corresponds to $(\sum_k F_k^i(t) - F_k^{i^{sched}}(t))$. The second component in the ACE is supposed to account for (1) the change in total power generation in an area needed to correct an offset in the (local) average frequency from a predetermined frequency set point (AGC function); and (2) the local load (demand) dependence on area frequency (LFC function). This response to frequency fluctuations is effected automatically by the governors of generators within an area, which compare an average frequency for the area to the desired set-point frequency calculated and fed to them by LFC/AGC.

---

[21] Depending on the actual type of system inputs and the level of noncompliance under competition, one may have a slower component $P_L[KT_t]$ also contributed to by some noncompliance with the scheduled transactions. The accounting for control services may be very sensitive to the accuracy of posing the load model.

The ACE signal for area i at time t is thus defined as

$$ACE^i(t) = b^i(\omega^i_{ave}(t) - \omega^i_{set}) + (F^i_{net}(t) - F^i_{sched}(t)) \qquad (3.34)$$

where $\omega^i_{ave}(t)$, ω^i_{set}, b^i, $F^i_{net}(t)$, and $F^i_{sched}(t)$ designate, respectively, the average frequency, the desired set-point frequency, the frequency bias scalar, the net sum of actual tie-line flows, and the net sum of the desired schedule of tie-line flows—all corresponding to area i.

One can show–assuming negligible transmission losses and steady-state system operation—that ACE^i for each area is equivalent to the generation/load power mismatch for that particular area [4]. As such, this signal may be interpreted as the amount of additional generation needed to return the area to its set-point frequency and scheduled power exchanges with other areas. A direct consequence of this relation is that each area is autonomous under ACE-based LFC/AGC in the sense that only in the area where a generation/load mismatch exists will the ACE be nonzero. Therefore, an LFC/AGC signal for area i resulting from ACE^i will activate changes in generation within area i only. The ACE is a quantity measurable by each area in a decentralized setting.

The best choice of the coefficient b^i and the most adequate performance criterion have been the main topic of several NERC committees [36]. This body, which includes representatives from utilities, establishes guidelines for all utilities. These guidelines, when met, are intended to realize the most reliable and economical operation of the interconnected system. Except for these guidelines, though, all LFC/AGC actions are nonmandatory at present.

Symbol b^i in (3.34) is known as the frequency bias scalar [94]. As indicated previously, for ACE^i to reflect generation/load mismatch in area i, this bias setting should directly reflect the net effect of (1) intra-area governor-adjusted generation, through governor actions, and (2) the load's response to frequency deviations from the frequency offset in area i. The trouble is that, at best, only a steady-state characterization of the automatic governor response at the individual governors in the area is known; moreover, load frequency estimates are typically inaccurate and thus, for all practical purposes, unknown [4]. Furthermore, one can argue that the simple summation of governor and load characteristics into a composite figure that does not account for electrical distances between the generators and loads within an area is not justifiable at present. In this sense, a direct relation between the steady-state parameters of individual generators and loads in the area, and the intra-area bias setting, b^i, is missing. In practice, b^i is often taken to be 1.5 percent of the total nominal load for the area. This inherent problem with the frequency bias setting is well recognized, and acknowledged by the inventor of LFC/AGC himself [1, 8].

Instead of fully reviewing the conventional setting of ACE-based generation and frequency control, we direct the reader to an excellent reference [4, pp. 1-10 and pp. 44-52]. This reference examines in depth the assumptions under which an ACE-type signal is a fully meaningful control signal. In contrast, fundamental issues concerning LFC/AGC when these assumptions are not valid are well documented, for example, in [2, 1, 91, 4, 6].

3.7.2 Dynamic performance objectives over long-term horizons in a horizontally structured industry

A closer look at the definition of ACE shows that this measurement signal is a combination of area frequency at each subsystem level and the net tie-line power flow exchange with the neighboring subsystems. As such, meeting performance criteria expressed in terms of the ACE (like criteria A1, A2 [36]) does not guarantee that the tie-line flow exchange by itself will be regulated to zero. Because of this, over long-term horizons one will observe potentially significant cumulative deviations in tie-line flows from their scheduled values. This phenomenon is referred to as inadvertent energy exchange (IEE) [1, 36]. To eliminate operating problems related to IEE, present practice has been to make up for these deviations from scheduled flows over long-term horizons by intentionally scheduling tie-line flow exchanges the next day so that each control area makes up for the IEE. There is no financial penalty associated with IEE at present, nor any mandatory specifications for IEE performance over long-term horizons.

3.7.3 Functional requirements for advanced LFC/AGC in a changing industry

Desirable functional specifications for an advanced LFC/AGC scheme might be summarized as follows:

- Advanced LFC/AGC should perform according to prespecified criteria important in a deregulated utility environment. Such criteria might include allowances for the presence of independent power producers within each utility, open access among different utilities, and incorporation of the large loads participating in demand-side management.

- The design of advanced LFC/AGC should preserve the simplicity of information exchange found in conventional LFC/AGC.

- Advanced LFC/AGC should coordinate economic criteria with an improvement in overall dynamic performance. This implies resolving the issue of what the frequency bias setting, b^i, should be.

- Advanced LFC/AGC should be sufficiently adaptive to accommodate unconventional energy management regimes. This should be done by accounting for the effects of unconventional entities on the overall dynamic performance of the interconnected system, and by allowing for on-line costing mechanisms that associate the value of generation and frequency regulation with each participant involved in systems control. The costing mechanisms should evolve from performance criteria that yield the best controlled performance of the interconnected system.

In assessing these functional requirements, we state several basic drawbacks in conventional LFC/AGC that prevent their realization.

3.7.4 Conceptual problems with meeting mid-term dynamic performance objectives by means of present AGC in a changing industry

Conventional ACE-type control schemes are based on several critical assumptions that we expect will be routinely violated in a market-driven utility environment:

- All the machines in an area are very closely electrically coupled relative to the coupling between any two areas [91].

- All machines participate in frequency regulation [91].

- There is no straightforward way of incorporating non-uniform machine characteristics within an area, such as differences between the damping coefficients, D_j^i, or speed regulation (droop) characteristics, σ_j^i. The frequency bias setting, b^i, is based on an aggregate droop characteristic for the area.

- Because of the above assumptions, the dynamic participation of different machines responding to an ACE signal cannot be easily defined.

- There are no fundamentals to associate the economic value of frequency control with the individual participants within an area.

- There are no means to assess the value of LFC/AGC for the case of open access among areas.

This list is by no means complete. However, it is a basis from which improvements to conventional LFC/AGC might spring. These suggested improvements, in fact, motivate the improved LFC/AGC scheme proposed in Chapter 5.

3.7.5 Conceptual problems with meeting long-term performance objectives in a changing industry

To analyze potential problems with meeting dynamic performance objectives over long-term horizons, such as on a half hour basis, or longer, we recall from

Section 3.1 that contracts may be basically of two types: long-term, firm and short-term, non-firm power inputs, responsive to price. In addition, both types of inputs could display noncompliance with their declared schedules, i.e., they could have random deviations of various duration and magnitude around the scheduled values. Also, deviations in native load, not necessarily price-driven, are superposed to the inputs.

Let us start by assuming that the long-term, firm transactions are accommodated for by static optimization tools such as OPF or its variations for meeting static performance criteria.

The dynamic performance objectives over mid- or long-term horizons now become matters to be handled by methods such as AGC or rules for managing IEE. It is particularly relevant in the context of meeting dynamic performance criteria to decide which level in the hierarchy is responsible for accommodating load deviations around the statically scheduled. At present it is assumed that sufficient regulation by means of AGC is available to regulate system frequency without causing transmission line constraints. Static tools such as OPF and ED are used instead to deal with the transmission constraints for anticipated demand.

If the system inputs consist of many short-term, non-firm scheduled transactions of significant magnitude relative to the magnitude of long-term, firm transactions present on the system, the preceding assumption is no longer valid. Because of this, it may become relevant, at the interconnected system level, to clearly state and enforce dynamic performance over long time horizons in response to short-term profit-driven inputs. The energy return as currently practiced in terms of IEE may not be adequate, since the participants may decide to game the system, returning IEE when energy is cheap, and using it when energy is expensive because of violations in operating constraints.

Particularly relevant will be the rules by which the operating constraints are managed in response to the actions of competitive market participants. It is possible to accommodate transactions at each subsystem level for the scheduled tie-line flows. This will result in a different system reliability and efficiency than sharing all the resources at the interconnected system level to accommodate the systemwide market, and to meet dynamic specifications on energy exchange over long time horizons.

It is essential to define and standardize these performance objectives. Otherwise, much gaming for profit will be seen among the market players actively responding to energy prices in a dynamic manner.

3.8 Static performance criteria for reactive power/voltage support

The choice of performance criteria optimal with respect to voltage regulation is at present unsettled, as discussed in Section 3.5.3. The static optimization tools are capable of computing optimal voltage once a performance criterion is chosen, but it remains an open question, particularly in a changing industry, as to how to define the notion of an optimal voltage profile for an interconnected system.

In deciding on rules for reactive power support and its value one should recognize the unique nature of reactive power propagation through an interconnected system. The reactive power changes are typically limited to the areas where reactive power imbalances are caused and can be eliminated to a great degree with systems control electrically close to the location causing problems [65]. Once this is recognized, one may come up with fairly straightforward rules for reactive power regulation. Recent work on finding the locations most affected by an input change in reactive power, and on finding the most effective systems control to eliminate the effect of the specific change, may be quite useful for this purpose [37]. On the other hand, it would be definitely wrong in the case of voltage problems to have zonal charges for simplicity, independent of the location causing the problem.

3.8.1 Criteria for mid- and long-term voltage control (AVC) at a subsystem level

Voltage regulation of power systems involves the excitation system to stabilize the generator terminal voltages to their given reference values. These reference values are adjusted at discrete instances slowly by the higher-level controls. Much effort has been given to analysis, modeling, and design of excitation systems [24]. In this text we do not discuss issues concerning the designs of fast primary controls. Instead, emphasis is on higher-level control design–the slow updating of the reference values for excitation systems. The goal is to develop reliable, automated regional and systemwide voltage control to enhance secure and economical operation of power systems. In the United States static optimization tools only are used to schedule the set-point voltages of generators for the anticipated reactive power demand.

Interesting examples of performance criteria on voltage changes over midterm horizons can be found in existing AVC schemes in France and Italy. The main objective of these schemes is to respond to the slow reactive load disturbances $Q_L[kT_s]$ defined in (3.32). The scheme is implemented at generator units whose voltage set points $V_G^{ref}[kT_s]$ are automatically changed to respond to deviations in load voltages $V_L[kT_s]$ at the chosen subset of loads, the critical "pilot point" loads, $V_c[kT_s] = C_s V_L[kT_s]$.

Similarly to the present AGC concepts, AVC at the secondary level should be designed in such a way as to keep operation of subsystems as autonomous as possible for given control constraints. Its main objectives are to

- Reschedule $V_G^{ref}[kT_s]$ at each subsystem level to meet reactive load deviations $Q_L[kT_s]$, $k = 0, 1, \ldots$.

- Regulate voltages $V_c[kT_s]$ to voltage values $V_c^{set}[KT_t]$ regulated at the highest system level.

- Regulate reactive tie-line power flow $F[KT_t] \equiv 0$ as long as reserves within each area are available (like the ACE principle for AGC).

- Optimize system performance (total reactive reserve or total transmission losses).

The performance specification in France for first-generation AVC was to have a controller in each region that would bring voltage deviations to the desired value in response to slow load changes within 3 minutes [103].

Second-generation AVC in France is even more interesting because it defines performance in terms of weighted relative quality of both relevant system outputs and the control cost for achieving it. Explicit expressions for the performance criteria of interest for voltage regulation at the subsystem and system levels are given in Chapter 6.

3.9 Summary

In this chapter, the performance criteria relevant for generation-based regulation of frequency and voltage are reviewed. At present, some of these criteria are recommended by NERC and are met by various utilities in a voluntary way. To retain the same level of power quality in a changing industry it will be essential to standardize these performance criteria and make them mandatory at various levels of the nested hierarchy. Participants in a competitive market, not concerned with meeting these criteria, will have to pay for systems control services needed to meet these objectives. If each subsystem has well-defined performance objectives, and if, in addition, the performance objectives are fully met up to the highest level of hierarchy, power quality under competition and nondiscriminatory valuing of systems control services can be assured. It is shown later in this text how the performance objectives defined in this chapter can be used to create a general operations framework under competition and relate technical services to the economic processes in such a way that the interconnected system approaches ideal performance. The same criteria are essential for allocating performance-based values of systems control to various market participants and charging for these.

One should not underestimate, however, the complexity of meeting the performance objectives in the real-time operations of a very large nested hierarchy. Because of this, it is essential to clearly assign performance objectives relevant at each level of hierarchy with minimal-order models. Our modeling framework, introduced in Chapter 4, offers promise in this direction, since it can be used without excessive new data for enhancing present systems control to meet performance criteria under competition.

Efficient generation-based system regulation of electric power systems in a competitive environment will require dynamic control schemes in addition to static optimization schemes because of excessive uncertainties in both generation and load. To implement dynamic optimization schemes, it is necessary to understand the basic meaning of the performance criteria chosen. This area has not been emphasized in the past in the power engineering literature because it was not needed.

Static optimization tools are based on a full information structure, or full state algorithms. This type of data structure is qualitatively different from the structure used at present for AGC and AVC, both of which are output feedback schemes that rely on a reduced information structure.

Generalizations in terms of reduced information structures of static optimization tools into optimal control tools over mid- and long-term horizons are essential both for good technical performance in a competitive market and for establishing performance-based value of systems control services. This is because static scheduling is not effective in dealing with the generation and load uncertainties typical of competitive management.

LFC/AGC performance has been deteriorating over recent years [92], because the conditions under which the LFC/AGC scheme was originally designed are often not met. A more dynamic mode now exists within electric power systems than was the case when LFC/AGC was first put into use [93]. Corrective actions are effected via mutual agreements among control areas—to keep tie-line power flow exchanges as scheduled and not to deviate excessively in the integral of frequency over time.

Typically, no financial penalties are assessed at present for deviations from scheduled power exchanges among areas, but this will have to change, particularly in a power industry that includes competition and open access. Based on problems experienced at present [93] it is believed that a more formal evaluation of the contributions of different areas to the overall technical and economic efficiency of the interconnected system will be necessary. The first step in this direction is to agree on the performance objectives that should be required in a changing industry.

New modes of energy management, such as open access and the presence of non-utility-owned generation in the system, as well as demand-side management of large industrial loads, require unconventional control solutions [21]. The need to establish a systematic control design free of restrictive operating assumptions is a prime motivation for our proposed approach. This approach should also provide adequate economic incentives to a variety of participants in new energy management schemes to adopt actions that support efficient operation of the interconnected system.

CHAPTER 4

STRUCTURAL MODELING AND CONTROL DESIGN USING INTERACTION VARIABLES

This chapter introduces a systematic, structure-based modeling framework for analysis and control of electric power systems for processes evolving over mid- and long-term time horizons. Much simpler models than the detailed dynamics for control design at different hierarchical levels are obtained by applying both temporal and spatial separation. The aggregate models represent the net effect of interactions among interconnected regions on specific hierarchical levels. They are exact, since no assumptions about weak interconnections among the subsystems are made. Moreover, they are easily understood in terms of power flows among the regions.

4.1 Structural modeling

This section introduces a new structure-based modeling approach to large electric power systems. While the approach recognizes the decomposition of the system into interconnected but administratively divided regions, it does not make any a priori assumptions with respect to strength of their interconnections. The administrative regions within the interconnected system are tied together through the tie lines, and the regional dynamics are coupled through the tie-line power flows. To maintain the traditional decentralized control structure, an administrative region is chosen as the base for study, for which dynamic models are derived explicitly in terms of the tie-line flows. It is shown that this framework of modeling captures fundamental properties of power system dynamics and facilitates the physical understanding of the inter-area dynamics.

Each administrative region consists of a certain number of generating and control units, and a transmission network that interconnects these generating and control units. Typically these units are located at different locations, and each individual generator has its own local control, in the sense that the control regulates output variables associated with this particular generator only. The schematic representation of the structure is shown in Figure 4.1.

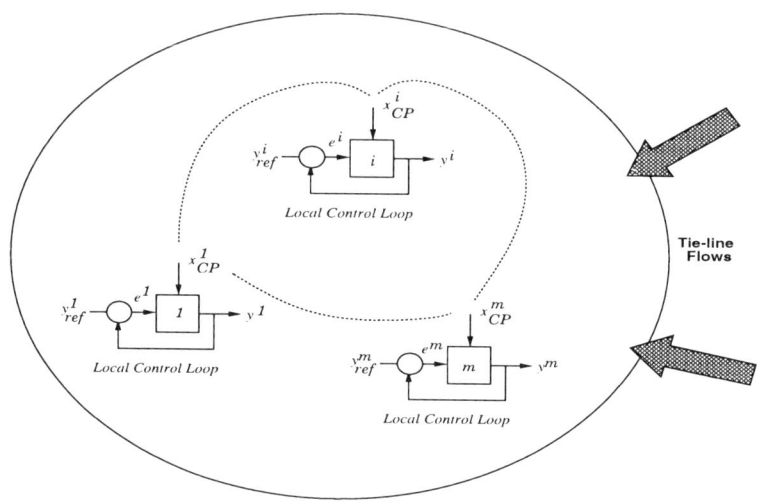

Figure 4.1: An administrative region with local controls

Dynamics of each individual unit, referred to here as the *local dynamics*, are derived in terms of local state variables of each unit. If the transmission network were not present, the local dynamics of these generating units would be completely decoupled. The transmission network constrains outputs of individual equipment by imposing power balance conditions on a subset of local variables of all generators and loads connected through the transmission network. The subset of local variables directly constrained by the system interconnections is referred to as the *coupling variables*. The local dynamics of individual units, in the form of ordinary differential equations (ODEs), together with the algebraic constraints imposed by the transmission network, form the dynamics of each administrative region in the form of differential-algebraic equations (DAEs).

It is well known that DAE problems are very difficult to study. The approach proposed here is to convert the DAE problem to an ODE problem by differentiating the network constraints under an assumption that holds for a very wide range of system operation. A standard state space nonlinear dynamic model for the administrative region in the form of unconstrained ODEs is obtained by combining local dynamics for individual units with the differentiated network constraints. The resulting dynamic model is in an extended state space, since the coupling variables also become states.

The developed structural dynamic model offers an essential, yet simple, vehicle for rigorous analysis of the power system dynamics. The model is uniquely suited for controlling inter-area dynamics at all time horizons. It also provides a theoretically solid basis for hierarchical control design of power systems to reject load disturbances over the widespread frequency spectrum as modeled in (3.1) and (3.3).

4.1.1 Modeling issues

Prior to introducing this modeling process, we review some general issues related to power system analysis and basic modeling assumptions, such as time scales, network and load modeling, the present control hierarchy, and the frequency/voltage decoupling assumption.[1]

Time scales

Modeling and control of large systems usually exploit the fact that various system processes take place at different rates of time and can thus be distinguished from each other. At different time scales, the dynamics and responses of the system exhibit different characteristics. Much of the analysis available at present is for studying the shortest response of the system to fast disturbances. Very little systematic analysis is available for system responses over mid- and long-term horizons. In order to partly eliminate this gap, it is accepted here that for short-term stable operations steady state outputs of fast dynamics can be viewed as moving equilibria under slower disturbances, forming a discrete event process (DEP) over a longer time scale. A fundamental difference between this class of processes and continuous dynamics is that a DEP under certain conditions in continuous dynamics is driven solely by control actions and disturbances.

Electric power systems, generally large in size and complex in operation, typically display this special class of processes. The local (primary) controllers stabilize system dynamics to within a threshold of their steady-state reference values with a very fast time constant. The steady-state outputs of these primary controllers are regulated at a regional (secondary) level with a significantly longer time constant than the primary controllers, forming what can be viewed as a DEP process. To fully optimize the operation of a system consisting of several electrically interconnected regions under varying load conditions, the reference values of the output variables from the secondary

---

[1] The modeling approach developed here is generalizable to the coupled dynamics.

level are adjusted at an even higher (tertiary) level with a still longer time constant.

Most important sources of different time scales include loads that vary over different time scales and different electrical distances within a large interconnected network. Power systems involve a huge number of devices interconnected over far geographical and electrical distances. The connections among these devices within an administrative region (electric power utility) are relatively meshed and strong, compared to very sparse and normally weak tie-line interconnections among different administrative regions. The meshed or strong intra-area connections represent shorter electrical distances, and the sparse or weak inter-area connections imply longer electrical distances. Loads typically have a widespread frequency spectrum. They are modeled as containing dynamics at three qualitatively different time scales: fast fluctuations, mid-term fluctuations, and long-term fluctuations. The controls responding at these three distinct time scales are the basis for a hierarchical control scheme to stabilize the frequency and voltage throughout the system.

As a convention throughout, we refer to the fast transient dynamics of the system as the *primary* process, with typical time constant T_p; the DEP on the mid-range time scale as the *secondary* process, with typical time constant T_s; and the slowest process as the *tertiary* process, with time scale T_t. The primary process is simply the continuous dynamics of the system, the secondary process is the set value adjustment by the regional controllers over the mid-term horizon, and the tertiary process is associated with the slowest adjustment of system settings relevant for the entire interconnected system.

Since the secondary and tertiary processes are activated only at discrete times, any variable v of interest can be decomposed into

$$v = v(t) + v[k] + v[K], \quad k, K = 0, 1, 2, \ldots \tag{4.1}$$

where $v(t)$ is the continuous component, associated with the primary dynamics with time scale T_p; the discrete secondary process is defined as $v[k] = v(kT_s)$; and $v[K] \stackrel{\Delta}{=} v(KT_t)$ is the slowest component, associated with the tertiary process. A schematic presentation of relevant time horizons is illustrated in Figure 3.4.

Network and load modeling

In studies of power systems, the transmission network is modeled as a linear circuit, normally with inductances, resistances, and shunt capacitances. For the frequency range of interest here, the dynamics of the transmission network are neglected. Under this modeling assumption, the transmission network effectively imposes algebraic constraints on the dynamics of the local generating units and their local controls. The constraints are simply the real and reactive power balances. Generator local dynamics in the form of

ODEs, together with the algebraic network constraints, lead to a constrained dynamic problem of a DAE form.

Loads are modeled as sinks or sources of real power P_L and reactive power Q_L. The deviations from constant power sinks/sources are further modeled as external disturbances to the system. Although more realistic models include dependence on their voltage and frequency, these models are not used at present for on-line control of the interconnected system. Using the notation introduced in (4.1), loads are represented by

$$P_L = P_L(t) + P_L[k] + P_L[K] \quad k, K = 0, 1, \ldots \qquad (4.2)$$

$$Q_L = Q_L(t) + Q_L[k] + Q_L[K] \quad k, K = 0, 1, \ldots \qquad (4.3)$$

for real power P_L and reactive power Q_L. In this equation, $P_L(t)$ and $Q_L(t)$ are the fast continuous fluctuations, $P_L[k]$ and $Q_L[k]$ represent variations over the mid-term horizon, and $P_L[K]$ and $Q_L[K]$ correspond to variations of the load on the long-term horizon.

Present control hierarchy

Corresponding to the three different time scales, the monitoring/control of large power systems typically employs a hierarchical structure:

- *Primary-level control*: This level is typically entirely localized in the sense that controllers respond only to changes in the local output variables. The main function of primary-level control is to correct for small, fast output deviations caused by fast load disturbances. Excitation and governor systems are the main primary controllers responsible for voltage and frequency controls, respectively.

- *Secondary-level control*: This level is concerned with changes at the regional level, considering interactions with the neighboring regions to be small. Its main function is to eliminate frequency and voltage deviations at certain critical locations over the mid-term horizon. These deviations are caused by slow load deviations. Adjusting the speed-changers of governors and the terminal voltages of generators are the main control objectives at this level.[2]

- *Tertiary-level control*: This level is concerned with the coordination of secondary-level controllers by incorporating the effects of interactions on the quasi-static changes of the interconnected system over the long-term horizon. The main goal of this coordination is to achieve systemwide optimal performance over the long-term horizon.

---

[2] The control hierarchy is described in terms of most standard controls, i.e., generator controls. Our concepts directly apply to many other controls available on modern power systems [73].

In terms of the load decomposition, the main purpose of the primary-level frequency and voltage controllers is to cancel the effects of the fast load fluctuations $P_L(t)$ and $Q_L(t)$. The regional secondary-level controllers are designed to control frequency and voltage under the slower load changes $P_L[k]$ and $Q_L[k]$. The systemwide tertiary-level controller is intended to update settings for the secondary-level controllers in response to slow nominal load changes $P_L[K]$ and $Q_L[K]$ over the time horizon T_t.

Decoupling assumption

It is well understood that in a static operation real power injections to a power system closely affect voltage phase angle differences across the transmission lines, and that reactive power injections closely affect voltage magnitudes, under normal operating conditions. In other words, sensitivities of real power to phase angle differences, and sensitivities of reactive power to voltage magnitudes are relatively larger than the cross-sensitivities of real power to voltage magnitudes and reactive power to phase angle differences. This property is referred to in the power systems literature as the real power/voltage decoupling assumption.

Under normal conditions, power systems operate in a quasi-steady state on slow time scales, i.e., the system reaches its steady state within the fast time scale T_p. Therefore, it is generally a valid assumption that real power/frequency dynamics and reactive power/voltage dynamics are decoupled in normal operating conditions. In order to exploit the most fundamental characteristics of power system dynamics, and not confuse their mathematics with the complex mathematical expressions associated with coupled dynamics, the decoupling assumption of real power/frequency and reactive power/voltage dynamics will be made throughout this text. This, however, does not imply that the decoupling assumption holds in general.

4.1.2 Modeling process

A complete dynamic model of any power system is a combination of the local dynamics and network coupling. This structural decomposition is schematically illustrated in Figure 4.1. For simplicity, we present the material using standard systems jargon instead of specialized power systems formulations in terms of frequency or voltage first. Specific illustrations of the results as applied to frequency and voltage regulation are described in Chapters 5 and 6 [16, 18, 74, 109].

4.1.3 Local dynamics

Let us define x_{LC}^i as the local states of generator i. One can express local dynamics of this generator in a general form as

$$\dot{x}_{LC}^i = f_{LC}^i(x_{LC}^i, u_{LC}^i, x_{CP}^i) \tag{4.4}$$

where $\dot{x}_{LC}^i$ is the derivative of local state with respect to time, u^i is the control input to the generator, and x_{CP}^i represents the coupling variables that relate the local dynamics of different generator sets. For example, in the case of real power/frequency dynamics, the coupling variable is simply the real power generation of the generator. The real power generation of different generators is coupled together by the transmission network that connects different generators. The primary task of a local controller is to stabilize the local output variables of a specific generator to their desired settings, which can be calculated locally or, more often, are supplied by higher-level controls. Assume that

$$y^i = C^i x_{LC}^i \tag{4.5}$$

is the vector of output variables of interest for generator i. Let y_{ref}^i represent the desired settings for output variables. The local control u_{LC}^i is typically a feedback control designed using the error signal

$$e^i = y^i - y_{ref}^i \tag{4.6}$$

After appropriate design of the local controller, the closed-loop dynamics of the generator set can be written as

$$\dot{x}_{LC}^i = f_{LC}^i(x_{LC}^i, y_{ref}^i, x_{CP}^i) \tag{4.7}$$

We now derive local dynamic models for all generator units in the network. As discussed previously, we choose any administratively divided region as the base system for our study. Consider here such a region consisting of m generator units. Define the local states, output variable settings, and coupling variables for the region as

$$x_{LC} \triangleq \begin{bmatrix} x_{LC}^1 \\ \vdots \\ x_{LC}^m \end{bmatrix}, \quad y_{ref} \triangleq \begin{bmatrix} y_{ref}^1 \\ \vdots \\ y_{ref}^m \end{bmatrix}, \quad x_{CP} \triangleq \begin{bmatrix} x_{CP}^1 \\ \vdots \\ x_{CP}^m \end{bmatrix} \tag{4.8}$$

Since (4.7) is true for any generator control set, one can simply obtain the local dynamic model for the entire region as

$$\dot{x}_{LC} = f_{LC}(x_{LC}, y_{ref}, x_{CP}) \tag{4.9}$$

where the nonlinear function is defined as

$$f_{LC}(x_{LC}, y_{ref}, x_{CP}) \triangleq \begin{bmatrix} f_{LC}^1(x_{LC}^1, y_{ref}^1, x_{CP}^1) \\ \vdots \\ f_{LC}^m(x_{LC}^m, y_{ref}^m, x_{CP}^m) \end{bmatrix} \qquad (4.10)$$

4.1.4 Network constraints

The transmission network imposes algebraic constraints on the coupling variables x_{CP} and a small subset of local state variables x_{LC}. With definition (4.8), it is demonstrated in [17] that the constraints are always given in the following structural form:

$$g(x_{CP}, x_{LC}, F) = 0 \qquad (4.11)$$

where the nonlinear function $g(\cdot, \cdot, \cdot)$ has the same dimension as that of the coupling variables. The term F represents tie-line flows into the region from its neighboring regions. This formulation assumes constant power loads. Any variations in the loads are viewed as disturbances to the system. The primary goal of power system control design is to cancel out the load disturbances. The differential equation of local dynamics given in (4.9), together with this algebraic relation, forms the dynamics of the system in the form of DAEs.

DAE problems are in general very difficult to handle, so we convert the DAE problem to ODEs by differentiating the algebraic constraint equation (4.11) with respect to time. It follows that

$$J_{CP}\dot{x}_{CP} + J_{LC}\dot{x}_{LC} + J_F \dot{F} = 0 \qquad (4.12)$$

where

$$J_{CP} \triangleq \frac{\partial g}{\partial x_{CP}}, \quad J_{LC} \triangleq \frac{\partial g}{\partial x_{LC}}, \quad \text{and} \quad J_F \triangleq \frac{\partial g}{\partial F} \qquad (4.13)$$

are defined as the Jacobian matrices of the network constraints. Note that these Jacobian matrices are evaluated at the actual value of the state and flow variables (x_{CP}, x_{LC}, F), and therefore no approximations are introduced.

To derive a standard state space ODE model for the dynamics of the region, let us assume that the square matrix J_{CP} is nonsingular. (If this is not the case, a complicated phenomenon such as impasse points [68] will occur. This case is outside the scope of this text and will not be further discussed.) Under the nonsingular condition, (4.12) can be equivalently written as

$$\dot{x}_{CP} = S_{LC}(x_{CP}, x_{LC}, F)\dot{x}_{LC} + S_F(x_{CP}, x_{LC}, F)\dot{F} \qquad (4.14)$$

where

$$S_{LC}(x_{CP}, x_{LC}, F) \triangleq -J_{CP}^{-1} J_{LC} \quad \text{and} \quad S_F(x_{CP}, x_{LC}, F) \triangleq -J_{CP}^{-1} J_F \qquad (4.15)$$

are defined as the sensitivity matrices of coupling variables to local states and flows, respectively. Again, these sensitivity matrices are functions of the state and flow variables, as explicitly indicated above. Eq. (4.14) represents an equivalent nonlinear ODE set for the network algebraic constraints.

4.1.5 Structural dynamic model

The local dynamic model (4.9), combined with the coupling dynamics given in (4.14), forms a complete set of ODEs for the dynamics of the specific region under consideration:

$$\begin{bmatrix} \dot{x}_{LC} \\ \dot{x}_{CP} \end{bmatrix} = \begin{bmatrix} f_{LC}(x_{LC}, y_{ref}, x_{CP}) \\ S_{LC}(x_{CP}, x_{LC}, F)\dot{x}_{LC} + S_F(x_{CP}, x_{LC}, F)\dot{F} \end{bmatrix} \quad (4.16)$$

or

$$\begin{bmatrix} \dot{x}_{LC} \\ \dot{x}_{CP} \end{bmatrix} = \begin{bmatrix} f_{LC}(x_{LC}, y_{ref}, x_{CP}) \\ S_{LC}(x_{CP}, x_{LC}, F)f_{LC}(x_{LC}, y_{ref}, x_{CP}) + S_F(x_{CP}, x_{LC}, F)\dot{F} \end{bmatrix} \quad (4.17)$$

Define the state variables in the extended state space for the region under study as

$$x \triangleq \begin{bmatrix} x_{LC} \\ x_{CP} \end{bmatrix} \quad (4.18)$$

and the nonlinear function on the right-hand side of (4.17) as

$$f(x, y_{ref}, F, \dot{F}) \triangleq \begin{bmatrix} f_{LC} \\ S_{LC}f_{LC} + S_F\dot{F} \end{bmatrix} \quad (4.19)$$

We obtain the nonlinear standard state space dynamic model for any administratively divided region as

$$\dot{x} = f(x, y_{ref}, F, \dot{F}) \quad (4.20)$$

Notice that the reference values for output variables, y_{ref}, are updated more slowly than the transient dynamics by a higher-level control center. The purpose of updating y_{ref} is to render an optimal performance of the system accommodating the slowly varying component of the load fluctuations (the fast component of the load variations is stabilized by the appropriate design of local controllers). This typical implementation of control, referred to in this text as the hierarchical control structure, gives rise naturally to different time scales for the closed-loop dynamics. The following section discusses the time scale separation associated with this particular structure.

Note also that this dynamic model is written explicitly in terms of tie-line flows into the region from neighboring systems. In (4.20), the tie-line flows act as an external input to the dynamics of the region under study. These flows play important roles in the inter-area behaviors of different regions within the interconnected system. It will be shown that the decoupled real power/frequency dynamics are not completely controllable under the present control structure; the local control u_{LC} cannot regulate inter-area behaviors of the interconnected system. The popular but not well-understood phenomenon of inter-area oscillations cannot be effectively suppressed with local governor controls without significantly changing voltages throughout the network. In this case, additional control actions are needed to guarantee the desired performance of the system. This leads to the idea of direct flow control described in Chapter 8, for which basic hardware is available at present [17].

4.1.6 Control-induced time scale separation

Because the updating of reference values of output variables is typically done more slowly than the time constant of the system transient dynamics, different time scales exist in the system dynamics over a long time horizon. Time scale separation techniques can be used to support higher-level control designs.

The linearized dynamic model for any administrative region is

$$\dot{x} = Ax + By_{ref} + UF + V\dot{F} \tag{4.21}$$

In this equation, vector F represents the tie-line flows into this region from its neighboring regions. The reference value y_{ref} is updated, by either the local level or a higher level, at discrete instants to regulate the profiles of output variables of direct interest so that some predefined optimality is achieved. Because of physical limitations and practical considerations, the updating is typically done more slowly than the transient dynamics. This process of updating the reference values of individual controllers is often called *secondary-level control* [103, 63].

Let us denote the time interval of secondary-level control as T_s, i.e., the reference value is updated at instant kT_s, $k = 0, 1, \ldots$. Thus the reference value y_{ref} is constant in the interval $kT_s < t < (k+1)T_s$. Let us further denote $v_s[k] = y_{ref}(kT_s)$ as a discrete time sequence of the reference value. With this notation, (4.21) can now be written as

$$\dot{x} = Ax + Bv_s[k] + UF + V\dot{F} \tag{4.22}$$

The objective of secondary-level control is to designate an appropriate discrete time sequence $v_s[k]$ to achieve some prespecified optimality criterion at each subsystem level.

Because the discrete sequence $v_s[k]$ varies more slowly than the transient dynamics, much simpler models can be derived to assist secondary-level control design. Let us now carry out the detailed derivations. Because the time constant of the transient dynamics is much shorter than the secondary-level control time interval T_s, one can assume that all transient dynamics settle to a steady state before each time instance kT_s, i.e., $\dot{x} = 0$ at kT_s. Eq. (4.22) then reduces to

$$Ax + Bv_s[k] + UF = 0, \quad \text{at } t = kT_s \tag{4.23}$$

or

$$Ax[k] + Bv_s[k] + UF[k] = 0 \tag{4.24}$$

using the convention of (4.1). Eq. (4.24) determines a static relation between the steady-state equilibria of the system and the reference values to be adjusted by secondary-level control.

The secondary-level control is designed to eliminate the slower steady-state offset of some critical variables in the region under the slow drifting of disturbances. Let us express these critical variables for the secondary level as

$$x_s = Dx \tag{4.25}$$

The dimension of x_s is in general much lower than the dimension of x. The reference value $v_s[k]$ is updated on the time scale T_s so that the slower steady state offset in x_s on the time scale T_s is eliminated. The desired relation between $x_s[k]$ and $v_s[k]$ can be determined from (4.24) as

$$x[k] = -A^{-1}Bv_s[k] - A^{-1}UF[k] \tag{4.26}$$

and therefore

$$x_s[k] = B_s v_s[k] + M_s(F[k] - D_s d_s[k]) \tag{4.27}$$

with $B_s \triangleq -DA^{-1}B$ and $M_s \triangleq -DA^{-1}U$.

Eq. (4.27) determines a quasi-static relation between $x_s[k]$ and $v_s[k]$. This quasi-static relation is best utilized for secondary-level control design when transformed into a dynamic model. To introduce the secondary-level discrete-time dynamic model, let us subtract (4.27) at two consecutive time instants kT_s and $(k+1)T_s$:

$$x_s[k+1] - x_s[k] = B_s(v_s[k+1] - v_s[k]) + M_s(F[k+1] - F[k]) \tag{4.28}$$

Define the update of the reference value, or the corrective control for the secondary level, as

$$u_s[k] = v_s[k+1] - v_s[k] \tag{4.29}$$

and the change of tie-line flows as

$$f[k] = F[k+1] - F[k] \qquad (4.30)$$

One obtains the secondary-level discrete-time dynamic model as

$$x_s[k+1] - x_s[k] = B_s u_s[k] + M_s f[k] \qquad (4.31)$$

Model (4.31) is introduced as the simplest model for designing output feedback-based secondary-level controllers at the regional level. This model can also be interpreted as representing a discrete event process of a moving equilibrium $x_s[k]$ driven by the discrete control actions $u_s[k]$ and the tie-line flows [67]. Variables $x_s[k]$ will be referred to as the *secondary-level states*.

It should be pointed out that the corrective control signal $u_s[k]$ defined in (4.29) represents an implicit integral control because, from (4.29),

$$v_s[l] = \sum_{k=0}^{l-1} u_s[k] + v_s[0] \qquad (4.32)$$

for any integer l. It is this implicit integral control that rejects the steady-state error in the output variables on the secondary-level time scale.

4.2 Hierarchical control design

In this section, we present the hierarchical control design methodology based on the foregoing time scale separation method. It is shown that the effects of neighboring regions can be easily accounted for in the regional control design, using the derived simple model at the secondary level.

4.2.1 Controllability

Let us first show a structural property associated with a control-driven system, i.e., the controllability of the system is determined by the relative dimensions of the states and controls. Assume that the dimension of the secondary-level states $x_s[k]$ is n, and the dimension of the secondary-level controls $u_s[k]$ is m. Recall that the controllability matrix of (4.31) [90], with $f[k]$ treated as an external input, can be written as

$$\begin{bmatrix} B_s & B_s & \cdots & B_s \end{bmatrix} \qquad (4.33)$$

This matrix has dimension $n \times nm$. If the number of controls m is less than the number of states n, as is typically the case, this controllability matrix is always rank-defficient, and the system is not fully controllable. This property is a structural one, since it is independent of the numerical values of the system.

As a result of this structural uncontrollability, only at most m states can be controlled independently. Let us choose m critical states as the output variables to be regulated by secondary-level control, expressed as

$$y_s[k] = C_s x_s[k] \qquad (4.34)$$

with matrix C_s having dimension $m \times n$. Variations in the output variables $y_s[k]$ can be easily obtained from (4.31) as

$$y_s[k+1] - y_s[k] = C_s B_s u_s[k] + C_s M_s f[k] \qquad (4.35)$$

Define the $m \times m$ square matrix $U_s = C_s B_s$. Then the above can be written as

$$y_s[k+1] - y_s[k] = U_s u_s[k] + C_s M_s f[k] \qquad (4.36)$$

Under the further assumption that the number of controls and the relevant outputs at each regional level are the same,[3] one obtains an explicit recursive relation between the controls, relevant output variables, and flows of the form

$$v_s[k+1] = v_s[k] + L_1 y_s[k+1] - L_2 f[k+1] + L_3[k] \qquad (4.37)$$

where

$$L_3[k] = -L_1 y_s[k] + L_2 f[k] + L_2 d_s[k] \qquad (4.38)$$

4.2.2 Conventional secondary-level control

The goal of secondary-level control is to stabilize the output variables $y_s[k]$ over the secondary time horizon to an optimal value determined by the tertiary control level. The conventional secondary-level control takes the simple proportional form

$$u_s[k] = G(y_s[k] - y_s[K]) \qquad (4.39)$$

where $y_s[K] \triangleq y_s(KT_t)$ is the optimal value for the output variables on the even longer tertiary time scale T_t. This value is calculated by the tertiary control level, and is constant for secondary processes.

Under this conventional feedback control, a secondary-level closed-loop dynamic model for output variables is obtained as

$$y_s[k+1] - y_s[k] = U_s G(y_s[k] - y_s[K]) + C_s M_s f[k] \qquad (4.40)$$

---
[3]This can be relaxed.

The gain matrix G can be chosen to optimize a performance index at the regional level

$$J_s = \sum_{k=0}^{\infty}(y_s^T[k]\,Q\,y_s[k] + u_s^T[k]\,R\,u_s[k]) \qquad (4.41)$$

for some matrices $Q = Q^T \geq 0$ and $R = R^T > 0$ specified by each region. The superscript ^{T} denotes the transpose of a matrix.[4] The optimization is with respect to $u_s[k]$, and the result is the optimal gain matrix G. In this process, tie-line flows with neighboring regions are neglected because of the large scale of the system and the desire to maintain the decentralized nature of regional control.

4.2.3 Improved secondary-level control

It is clear from (4.40) that tie-line flows viewed as an independent external input to the system affect the dynamics of the output variables. The conventional optimal control (designed neglecting interconnections) will no longer be optimal when implemented in an actual system where interconnections are indeed present. To fully compensate for the effect of interconnections, we propose a modified feedback control law in the form

$$u_s[k] = G(y_s[k] - y_s[K]) + Hf[k] \qquad (4.42)$$

where the term $Hf[k]$ is to cancel the effect of $f[k]$ on output variables. Substituting (4.42) into (4.40) yields

$$y_s[k+1] - y_s[k] = U_sG(y_s[k] - y_s[K]) + (U_sH + C_sM_s)f[k] \qquad (4.43)$$

It is clear that if U_s is invertible, then the effects of the tie-line flows can be fully eliminated by simply choosing

$$H = -U_s^{-1}C_sM_s \qquad (4.44)$$

With this choice of H, (4.43) reads

$$y_s[k+1] - y_s[k] = U_sK_s(y_s[k] - y_s[K]) \qquad (4.45)$$

with no flows entering into the equation. In other words, the region under study looks as if it were fully isolated from the rest of the system as far as the output variables are concerned.

Note that the condition that U_s is invertible should not be viewed as restrictive; instead, it ought to be taken as one of the requirements for the

---

[4]Both secondary- and tertiary-level controls are regulator functions that respond to deviations from scheduled outputs to meet anticipated system changes.

choice of output variables. The matrix $(I+U_s K_s)$ is the system matrix for the output variables $y_s[k]$ seen from (4.40) or (4.45); therefore, if the matrix U_s were singular, the closed-loop system matrix $(I + U_s K_s)$ would always have an eigenvalue of 1 and consequently steady-state errors would be inevitable for the chosen output variables. To fully control all output variables, it is required that they be selected such that U_s is of full rank.

Note also that the control scheme presented here is totally decentralized, assuming that tie-line flows are locally measurable at each region level. No detailed information about neighboring regions is needed; only tie-line flows are required, since they aggregate the net effect of detailed dynamics of neighboring regions. It is not an unrealistic assumption that tie-line flows are locally measurable.

4.2.4 Quasi-static interaction variables

The secondary-level quasi-static dynamic model for any region explicitly in terms of the tie-line flows has been derived as (4.28), where $x_s[k]$ is the state vector representing all buses in the region. The dimension of the sensitivity matrix B_s is $n \times m$, with n being the total number of buses in the region, and m the number of generator buses that participate in secondary-level control. In general it is true that $n > m$, i.e., the number of load buses is larger than the number of generator buses participating in secondary-level control.

Under the condition of $n > m$, one can easily verify that the closed-loop system using any feedback control is singular, because matrix B_s has maximum rank of m. This structural singularity is due to the relative numbers of controls and states. This is a general property for any control-driven system. In exploiting this structural singularity of quasi-static voltage dynamics, we first give the following definition.

- **4.1 Definition (Quasi-Static Interaction Variables):** *Any linear combination of the states, $z[k] = Tx[k]$, $T \neq 0$, that satisfies*

$$z[k+1] - z[k] \equiv 0, \quad \forall k \tag{4.46}$$

  for any secondary-level control actions, and in the absence of interactions among regions and the disturbance, i.e., $f = 0$ and $d_s = 0$, is defined as the quasi-static interaction variable of the administrative region under study.

The quasi-static interaction variables do not vary with time when interconnections are removed and load disturbances are not present. For the interconnected system, therefore, any variations of the interaction variables with time are entirely due to the interactions among regions or load disturbances. It should be noted from the definition that the interaction variables are not unique. In fact, any combination of the interaction variables is still a set of interaction variables.

Let us derive the condition for the transformation matrix T. Combining (4.46) and (4.28) yields

$$z[k+1] - z[k] = TB_s u_s[k] + TD_s(f[k] - d_s[k]) \qquad (4.47)$$

Under the conditions in the definition, $f[k] \equiv 0$ and $d_s[k] \equiv 0$, we arrive at

$$z[k+1] - z[k] = TB_s u_s[k] \qquad (4.48)$$

In order to have $z[k+1] - z[k] \equiv 0$ for any control $u_s[k]$, matrix T must satisfy

$$TB_s = 0 \qquad (4.49)$$

This is the desired equation for calculating T. Note that matrix B_s has maximum rank $m < n$, and therefore (4.49) has nonzero solutions for T. It is quite easy to solve T from (4.49), since it is a simple algebraic equation and can be solved using a Gaussian elimination method. The need for eigenstructure analysis is completely avoided.

Note that the definition for interaction variables is for any secondary-level control, meaning that the interaction variables are independent of specific control. Equivalently, secondary-level control cannot affect the interaction variables. Any variations of the interaction variables are uniquely due to the interactions with other regions or the load variations. The matrix T, as a result, will not be dependent on the specific form of secondary-level control.

Once the interaction variables are determined from (4.49), one can further derive the dynamic model for these interaction variables. Eqs. (4.47) and (4.49) simply lead to

$$z[k+1] - z[k] = TD_s(f[k] - d_s[k]) \qquad (4.50)$$

This is the desired dynamic model for the interaction variables. This simple model relates the interaction variables to the tie-line flows and load variations. It is of crucial importance for secondary-level control and tertiary-level coordination, as is discussed in more detail later.

Notice that the definition for interaction variables does not assume numerically weak interconnections. Rather, it reflects a structural property of the system, different numbers of the states and controls. It is interesting to relate the interaction variables defined above to the slow variables in singular perturbation analysis when the interconnections are indeed weak. It is seen from the interaction dynamic model (4.50) that, in the weak interconnection case, the interaction variables do vary more slowly than the rest of the states. One can rigorously prove that, in the weak interconnection case, the interaction dynamics derived here will be the slow subsystem in the singular perturbation analysis.

For illustration of the interaction variables, see [17]. To summarize, this section presents a structure-based modeling and control approach for dynamics of an interconnected power system. Dynamics of the system are formulated by combining the local dynamics of individual generators and the network couplings. Quasi-static dynamic models on slower time scales are derived. The structural models developed here have been used to define systemwide voltage control on slower time scales according to performance specifications at each regional level [16, 109]. Both conventional and improved secondary-level voltage control are studied using this framework.

4.3 Tertiary level coordination

With an increased tendency toward large energy transfers over far distances, the problem of maintaining voltages and frequency within acceptable operating specifications has emerged in the operation and planning of power systems throughout the world.

The main purpose of tertiary-level controls is to update set values for tie-line power flows $F[K]$, $K = 0, 1, \ldots$ on the tertiary-level time scale in order to optimize systemwide performance for the anticipated base load $P_L[K]$ and $Q_L[K]$, $K = 0, 1, \ldots$. This could be done on an hourly basis, if not more often, in accordance with the statistical information on base load. The actual setting of tie-line flows is achieved by changing the settings of secondary-level controllers. Because this is done so infrequently, it could involve recomputing basic matrices around a new operating point for the anticipated load over the time horizon T_t.

The optimization problem for the optimal set values can be formulated for three basic architectures (industry structures):

- *Fully centralized*: The coordination tasks are performed by a systemwide coordination center that has full information. The entire interconnected system is modeled as a single region. One single performance criterion is optimized, and the optimal set values for all regions are obtained.

- *Fully decentralized*: The determination of set values of the controllers is done by each individual region itself. Each region optimizes its own performance criterion. Each region does not assume any information about the rest of the system. In the optimization process of each region, the tie-line flows into the region are measured and used to determine the optimal pilot voltage set values for the region.

- *Combined centralized/decentralized*: In this scheme, each region assumes limited information about the rest of the system, and with the limited information, tries to optimize its own performance criterion. The natural choice for the limited information about the rest of the system is simply the aggregate model developed here. We model this scenario in a game theoretical setting.

The systemwide performance criterion for the fully centralized method is in general quite difficult to establish, and the computational effort for the solution is enormous because the power system is very large. Therefore, the coordination scheme cannot be implemented often. The fully centralized method also requires systemwide communication over far distances. The fully decentralized or partially centralized/decentralized schemes, on the other hand, have obvious advantages. Specific problems of different regions can be handled with different performance criteria. This is particularly suitable for a multi-utility environment. The performance criteria for smaller regions are easier to obtain, and the computational work is significantly reduced. As a result, no systemwide communication is required, and the coordination schemes can be implemented relatively more frequently. The major disadvantage of the scheme under a competition market is that instability can occur.

Because of the large size of the system and the complexity of system operation, any practical on-line coordination schemes must be based on a reduced-information structure in order to be efficiently applied. In this text, we develop a tertiary-level coordination scheme based on a reduced information structure using the interaction variables defined previously. The defined interaction variables represent inter-area tie-line power flows and serve as a basis for inter-area coordination. We derive an important relation between the quasi-static interaction variables and the set values of critical outputs at each regional level.

4.4 New tertiary-level aggregate model

Using our structural modeling approach, we can derive the relation between the critical subsystem outputs at each regional level and the regional controls, the relation between the flows and the interaction variables, and the relation between the flows and the output variables, on the tertiary-level time scale T_t. These relations serve as constraints to the optimization problem for determining optimal set values for the most relevant output variables. We describe here only the most relevant relations in the context of this text; for others, see [17].

In this section, the whole interconnected system is considered as one single region. Since this big single region is an isolated system, there are no tie-line flows into the system. All previous derivations carry over to the whole interconnected system, except all tie-line flow terms drop out.

We use boldface letters to represent any variable associated with the whole interconnected system. For example, let us define $\mathbf{v}_s$, $\mathbf{y}_s$, and $\mathbf{d}_t$ to represent control, the critical output variables, and the load disturbances of the whole system. The relation between the critical outputs and controls is in the same form as (4.37) and (4.38):

$$\mathbf{v}_s[K+1] = \mathbf{L}_1 \mathbf{y}_s[K+1] + \mathbf{L}_3[K] \tag{4.51}$$

where

$$\mathbf{v}_s[K] = \mathbf{L}_3[K] - \mathbf{L}_1 \mathbf{y}_s[K] + \mathbf{L}_2 \mathbf{d}_t[K] \tag{4.52}$$

Comparing with (4.37) and (4.38), we observe the absence of the flow term, because the whole system is assumed to be an isolated one, and there is no flow for the isolated system.

The relation between the internal flows among the regions within the interconnected system and the critical output variables is obtained by combining (4.51)–(4.52) with (4.37) and (4.38) written for all regions. Its general form is

$$\mathbf{S}(\mathbf{F}[K+1] - \mathbf{F}[K]) = \mathbf{L}(\mathbf{y}_s[K+1] - \mathbf{y}_s[K]) + \mathbf{L}_d \mathbf{d}_t[K] \tag{4.53}$$

This model is referred to as the *tertiary-level model*, which explicitly relates the interaction variables among the regions and the critical output variables in the interconnected system. One should observe a difference between this model and the model expressed explicitly in terms of flows. The latter is not possible in the case of a general power system, implying that at the tertiary level only interaction variables, and not individual tie-line flows, can be set explicitly.

4.5 Comparison of the proposed control structures to those used at present

The only two closed-loop schemes working at present are automatic generation control (AGC) regulation schemes at the subsystem level of horizontally structured power systems and secondary automatic voltage control (AVC) in France and Italy. As described in Chapter 3, AGC uses a single scalar measurement known as the area control error (ACE). As such, it only represents power mismatch as seen at the subsystem level, and it is not capable of accounting for the effects of non-utility-owned generation inside the control area.

AVC, on the other hand, uses several measurements of critical ("pilot") voltages within a subsystem. This has mainly been the case in order to account for localized response to reactive power/voltage changes.

In some sense, this scheme has served as our motivation to develop control structures for both frequency and voltage regulation that are capable of incorporating non-utility-owned controls, when desired. In the case of AGC, this has required generalizing measurement from a scalar-valued ACE to the vector of measurements.

At present, the tertiary-level coordination has not been automated anywhere in the world.

4.6 Summary

This chapter provides a structure-based approach to modeling and control design for large electric power systems using minimal information. Starting with an interconnected system as a whole, models are derived in terms of state variables relevant for the hierarchy level of interest and the interaction variables between this level and the remaining system. The unique attribute of the mathematics presented here is in terms of the time scales considered. Traditionally, electric power system models for power systems have been used to analyze very short-term responses (transients) or static operation. It turns out that the most interesting and relevant processes are those involving moving equilibria over long time horizons, in which system input changes are driven by slow load variations from their expected values or by price-responsive consumers. Dynamic models are essential for studying relations between economic processes and technical processes, including the systems control needed over long time horizons to keep the system together.

The long time scales are particularly pronounced at the secondary and tertiary system levels. Our theoretical framework applies to the existing horizontal system structure as well as to architectures that are likely to evolve under competition.

Our modeling and control framework provides models at different hierarchy levels. It is useful for formulating processes under competition, and it takes into account the various performance objectives of different levels yet is consistent with systemwide performance objectives. This is because the lower-level model, which can be thought of as representing individual market participants, is expressed in terms of local states and controls associated with this level and in terms of *interaction variables* with the other participants.

Distributed decision making, as shown in Chapter 7, takes place without any tertiary-level coordination. The problem of technical performance becomes one of fully distributed controls at the secondary level with individual performance objectives. With our approach it is possible to set up different decision-making processes by means of secondary-level models in order to study the stability of the interconnected system over long time horizons. More important, one can study the optimality of the interconnected system as a function of the level of decentralization, i.e., as a function of the type of industry structure.

The tertiary-level model is essential for efficient use of generation-based systems control at the interconnected level. The model is unique relative to previously derived aggregate models because it is expressed in terms of physically measurable interaction variables for the whole system. The model is relatively low-order, since it reflects only the relation between the interaction variables and the controls designated for regulation. The intra-area variables at the lower levels of hierarchy do not enter into this model.

Potential applications of the derived models are illustrated throughout this text.

CHAPTER 5
GENERATION-BASED REGULATION OF REAL POWER/FREQUENCY

5.1 State of the art and potential problems of frequency regulation

This chapter concerns the regulation of generation and frequency in large electric power systems. Its main objective is to correct for a net real power supply/demand mismatch resulting from fluctuations in demand around the anticipated. These deviations from anticipated demand are compensated for by scheduling generation in the order of merit according to the economic dispatch/unit commitment methods.

As the industry structure changes, it is important to have well-understood performance criteria and the means of achieving them for systematic real power/frequency regulation. The supply/demand mismatch is caused by fluctuations in native load and by a variety of scheduled economic transactions and noncompliance with these transactions. While the effects of changes in native load are at present fairly well understood, the dynamics of real power supply/demand imbalances caused by a mix of native load fluctuations and economic transactions are not. It is necessary to clarify projections of imbalances over mid- and long-term horizons and the regulation of generation-based real power according to prespecified performance criteria, as discussed in Chapter 3.

This situation calls for dynamic regulation instead of using static optimization tools such as optimal power flow and economic dispatch under the assumption that the supply/demand mismatch is certain. The premise here

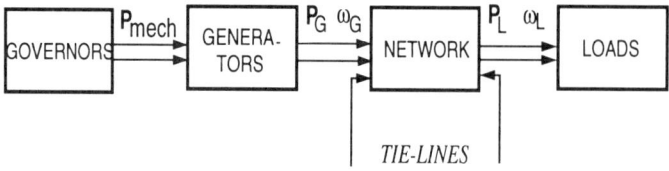

Figure 5.1: Typical structure of real power/frequency control

is that the performance criteria expressed in Section 3.7.1, such as A1 and A2, can only be met by means of systematic dynamic regulation. It is no longer sufficient to rely on static scheduling methods.

Recall the performance criteria for nested, large structures, described in Chapter 2. Two qualitatively different scenarios are envisioned: (a) when ex post charges for real power regulation are accepted without an active adaptation of system inputs to these charges, and (b) when real-time economic signals are given to the parties causing the supply/demand mismatch. The latter leads to potentially more efficient energy management than the former. However, both (a) and (b) are challenges to existing real power regulation, in terms of meeting both the technical and the economic objectives.

In this chapter, a systematic model in terms of the interaction variables presented in Chapter 4, applicable to the nested hierarchy described in Chapter 2, is described in the context of real power/frequency regulation in a changing industry. In practice, conventional frequency regulation is automated and implemented in separate administrative divisions within a large electrically interconnected system.

5.2 New modeling

In this section, the general structure-based modeling approach proposed in Chapter 4 is used for modeling real power/frequency dynamics. Load disturbances are explicitly modeled.

5.2.1 Local dynamics

The typical structure for real power/frequency control is shown in Figure 5.1. The governor control regulates the mechanical power applied to the generator shaft to stabilize the frequency. At present, the control structure is entirely localized in the sense that each governor control uses only the frequency error signal, defined as the difference of the measured frequency and the given reference value, of the particular generator to which it is connected. Direct coordination among the controls of different generator units does not exist.

Let us first study the local dynamics of each generator unit. Consider any generator unit i. Since the generator is simply a rotating shaft, the dynamic model is simply the mechanical rotation equation

$$J^i \dot{\omega}^i = \tau_m^i - \tau_e^i - \tau_d^i \tag{5.1}$$

where ω^i is the rotating speed, or frequency, of the generator, and J^i is the inertia of the rotating shaft. The terms τ_m^i, τ_e^i, and τ_d^i represent the input mechanical torque, output electromagnetic torque, and mechanical damping torque, respectively. To simplify notation, the superscript i will be dropped throughout this section. Eq. (5.1) then becomes

$$J\dot{\omega} = \tau_m - \tau_e - \tau_d \tag{5.2}$$

This equation is typically converted into a more convenient form using power instead of torque. Multiply (5.2) by ω. Recognizing that multiplication of torque with frequency is power, this leads to

$$M\dot{\omega} = P_m - P_e - P_d \tag{5.3}$$

where $M = \omega J$, and P_m, P_e, and P_d are the input mechanical power, output electric power, and mechanical damping power. Since the generator is operating very close to the nominal frequency ω_0 (60 Hz in the United States), one typically takes $M = \omega_0 J$ as a constant. The damping power P_d is usually small and is assumed to be a linear function of the frequency, $P_d = D\omega$, where D is the damping coefficient. For the linear damping, (5.3) becomes

$$M\dot{\omega} + D\omega = P_m - P_e \tag{5.4}$$

The mechanical power P_m is regulated by the governor control.

Consider a simple governor-turbine-generator (G-T-G) set, shown in Figure 5.2. The governor regulates the valve opening a of the turbine, which in turn controls the mechanical power applied to the generator shaft P_m. We adopt the same notation as in [3]. Turbine and governor are in general modeled as a first-order system, given by

$$T_u \dot{P}_m = n(P_m, a) \tag{5.5}$$

$$T_g \dot{a} = m(a, \omega^{err}) \tag{5.6}$$

where T_u and T_g are the turbine and governor time constants. The frequency error signal is defined as $\omega^{err} = \omega - \omega^{ref}$, with ω_G^{ref} being the reference value for the governor.

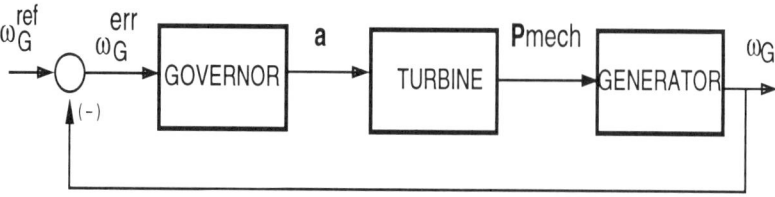

Figure 5.2: Primary control loop of a G-T-G set

The local state variables of the G-T-G set are defined as the frequency, mechanical power, and valve position, given by

$$x_{LC} \triangleq \begin{bmatrix} \omega \\ P_m \\ a \end{bmatrix} \tag{5.7}$$

The local dynamic model of the G-T-G set is described in nonlinear state space form:

$$\dot{x}_{LC} = f_{LC}(x_{LC}, \omega^{ref}) - cP_e \tag{5.8}$$

where f_{LC} is the combination of (5.4)–(5.6), and the vector c is given by

$$c = \frac{1}{M} \begin{bmatrix} 1 \\ 0 \\ 0 \end{bmatrix} \tag{5.9}$$

Assume that there exists a nominal operating point given by $x_{LC}^0 = (\omega^0, P_m^0, a^0)$, and $P_e = P_e^0$. Define deviations of the local state variables from the nominal operating point as

$$\begin{aligned} \delta\omega &\triangleq \omega - \omega^0 \\ \delta P_m &\triangleq P_m - P_m^0 \\ \delta a &\triangleq a - a^0 \end{aligned} \tag{5.10}$$

or $\delta x_{LC} \triangleq x_{LC} - x_{LC}^0$. The linearized local dynamics can be derived as

$$\delta\dot{x}_{LC} = A_{LC}\delta x_{LC} + b\delta\omega_G^{ref} - c\,\delta P_e \tag{5.11}$$

where A_{LC} is the system matrix of local dynamics of each generator, and given by

$$A_{LC} = \begin{bmatrix} -D/M & 1/M & e_T/M \\ 0 & -1/T_u & K_t/T_u \\ -1/T_g & 0 & -r/T_g \end{bmatrix} \tag{5.12}$$

Chapter 5

$$b = \frac{1}{T_g} \begin{bmatrix} 0 \\ 0 \\ 1 \end{bmatrix} \tag{5.13}$$

Quantities T_u, K_t, T_g, and r are constant parameters (see [3] for more details). To simplify notation, the prefix δ will be omitted throughout this section. With this convention, (5.11) becomes

$$\dot{x}_{LC} = A_{LC} x_{LC} + b\omega_G^{ref} - cP_e \tag{5.14}$$

Now let us write the local dynamics for all generator units in the region. Eq. (5.14) is true for any generator, i.e.,

$$\dot{x}_{LC}^i = A_{LC}^i x_{LC}^i + b^i \omega_G^{i\,ref} - c^i P_e^i, \quad i = 1, \ldots, m \tag{5.15}$$

Define the local state, frequency reference, and generator output for the region as

$$x_{LC} \triangleq \begin{bmatrix} x_{LC}^1 \\ \vdots \\ x_{LC}^m \end{bmatrix}, \quad \omega_{ref} \triangleq \begin{bmatrix} \omega_{ref}^1 \\ \vdots \\ \omega_{ref}^m \end{bmatrix}, \quad P_G \triangleq \begin{bmatrix} P_e^1 \\ \vdots \\ P_e^m \end{bmatrix} \tag{5.16}$$

and the regional local system matrix

$$A_{LC} = \begin{bmatrix} A_{LC}^1 & & \\ & \ddots & \\ & & A_{LC}^m \end{bmatrix} \tag{5.17}$$

and

$$b \triangleq \begin{bmatrix} b^1 & & \\ & \ddots & \\ & & b^m \end{bmatrix}, \quad c \triangleq \begin{bmatrix} c^1 & & \\ & \ddots & \\ & & c^m \end{bmatrix} \tag{5.18}$$

One obtains the local dynamic model for the entire region as

$$\dot{x}_{LC} = A_{LC} x_{LC} + b\omega_G^{ref} - cP_G \tag{5.19}$$

The vector of generator power outputs P_G is the coupling variables x_{CP} introduced in Chapter 4, which serve as the link to local dynamics of other generators via the transmission network.

5.2.2 Network coupling

Consider any single region with m generators. Network constraints are typically expressed in terms of nodal type equations that require complex valued power into the network $\hat{S}^N$ to be equal to the complex valued power $\hat{S} = P + jQ$ injected into each node:

$$\hat{S}^N = \hat{S} \tag{5.20}$$

where $\hat{S}^N = P^N + jQ^N$ is the vector of net complex power injections to all nodes, which can be written as

$$\hat{S}^N = \text{diag}(\hat{V})\hat{\mathbf{Y}}_{bus}^* \hat{V}^* \tag{5.21}$$

where $\hat{\mathbf{Y}}_{bus}$ is the admittance matrix of the network, $\hat{V} = Ve^{j\delta}$ is the vector of all nodal voltage phasors, with magnitude V and phase δ. The notation diag(.) stands as the diagonal matrix with each element of the vector as the diagonal element. Here we focus only on the real power constraints. The constraints for reactive power are discussed in Chapter 6. The real part of this equation becomes

$$P^N = P^N(\delta, V) \tag{5.22}$$

Further, the real power injected into each generator terminal on the interconnected system, P, generally is the sum of the generator power output P_G and the real power flow from the neighboring areas. Let us define the vector of the real power flow from the neighboring areas into all generator nodes as

$$F_G \triangleq \begin{bmatrix} F_G^1 \\ \vdots \\ F_G^m \end{bmatrix} \tag{5.23}$$

Then it is obvious that $P_G^N = F_G + P_G$. The network constraints for the real power balance can be further written as

$$P_G^N(\delta, V) = F_G + P_G \tag{5.24}$$

Similarly, since the real power from the load into the network can be written as the difference of real power flow injected into the network at the load terminal F_L and the real power absorbed by the load P_L, i.e.,

$$P_L^N = F_L - P_L \tag{5.25}$$

the network constraints (5.22) at the load nodes are expressed as

$$P_L^N(\delta, V) = F_L - P_L \tag{5.26}$$

Chapter 5

assuming that the positive direction for tie-line flows is the injection into the network, and the positive direction for loads is leaving the network. The separation of the power injection into the part from the actual device and the part from the tie lines with the neighboring systems is essential for establishing structural models of the interconnected system.

Differentiation under the decoupling assumption ($\partial P^N / \partial V = 0$) yields

$$F_G + P_G = J_{GG}\delta_G + J_{GL}\delta_L \tag{5.27}$$

$$F_L - P_L = J_{LG}\delta_G + J_{LL}\delta_L \tag{5.28}$$

where

$$J_{ij} = \frac{\partial P_i^N}{\partial \delta_i}, \quad i,j = G, L \tag{5.29}$$

are the Jacobian matrices evaluated at the given equilibrium operating point. Assuming J_{LL} to be invertible under normal operating conditions, we define one of the most important matrices associated with a transmission network, the sensitivity matrix,

$$C_\omega \triangleq -J_{LL}^{-1} J_{LG} \tag{5.30}$$

to express frequency deviations at loads ω_L in terms of frequency deviations at generators ω_G and fluctuations in load power. It follows from (5.28) that

$$\delta_L = C_\omega \delta_G + J_{LL}^{-1}(F_L - P_L) \tag{5.31}$$

or by differentiating with respect to time,

$$\omega_L = C_\omega \omega_G + J_{LL}^{-1}(\dot{F}_L - \dot{P}_L) \tag{5.32}$$

where

$$\omega_G = \dot{\delta}_G \triangleq \begin{bmatrix} \omega^1 \\ \vdots \\ \omega^m \end{bmatrix} \tag{5.33}$$

Eq. (5.32) defines the explicit dependence of load frequencies on generator frequencies determined by the network constraints. Combining (5.31) and (5.27), and defining the other two most important matrices associated with a transmission network,

$$K_P \triangleq J_{GG} + J_{GL} C_\omega \tag{5.34}$$

and

$$D_P \triangleq -J_{GL}J_{LL}^{-1} \tag{5.35}$$

results in

$$P_G = K_P \delta_G + D_P P_L - F_e \tag{5.36}$$

Here F_e represents effective tie-line flow as seen by each generator and is given as

$$F_e \triangleq F_G + D_P F_L \tag{5.37}$$

It follows after taking the derivative with respect to time on both sides of (5.36) that

$$\dot{P}_G = K_P \omega_G - \dot{F}_e + D_P \dot{P}_L \tag{5.38}$$

This equation defines the relation among all the generator real power outputs P_G, the tie-line flows into the system, and the load variations through the network characteristics specified by the two important matrices K_P and D_P.

It should be noted that any (portion of a) network is fully characterized by the three matrices (K_P, C_ω, D_P), with K_P reflecting the effect of the generator frequencies on the generator real power outputs, C_ω relating the generator frequencies to the load frequencies, and D_P representing different electrical distances of loads at different locations seen by the generators. The structural properties of these three matrices determine the fundamental features of the power system's primary real power/frequency dynamics. They have a direct impact on important issues such as the inter-area oscillations. These properties also have a systemwide effect on the higher-level dynamics. Detailed studies on these matrices are presented in the next section.

5.2.3 Regional dynamics

The state space formulation of the linearized local dynamics of all the governor control sets on the system (5.19), together with the network constraint equation (5.38), forms the closed-loop dynamic model of the interconnected system. To derive this, let us first repeat (5.38) here:

$$\dot{P}_G = K_P \omega_G - \dot{F}_e + D_P \dot{P}_L \tag{5.39}$$

Generator frequencies are part of the local generator states, given by

$$\omega_G = E x_{LC} \tag{5.40}$$

with the matrix

$$E = \text{BlockDiag}(e^1, \ldots, e^m) = \begin{bmatrix} e^1 & & \\ & \ddots & \\ & & e^m \end{bmatrix} \quad (5.41)$$

and $e^i = \begin{bmatrix} 1 & 0 & \cdots & 0 \end{bmatrix}^i$, with the dimension matching the dimension of the local states x_{LC}^i, $i = 1, \cdots, m$. Local dynamics are derived in (5.19). Combining these two equations, the standard state space linearized model of the single region within the interconnected system in terms of the tie-line flows explicitly takes the form

$$\begin{bmatrix} \dot{x}_{LC} \\ \dot{P}_G \end{bmatrix} = \begin{bmatrix} A_{LC} & -c \\ K_P E & 0 \end{bmatrix} \begin{bmatrix} x_{LC} \\ P_G \end{bmatrix} - \begin{bmatrix} 0 \\ \dot{F}_e \end{bmatrix} + \begin{bmatrix} 0 \\ D_P \end{bmatrix} \dot{P}_L \quad (5.42)$$

The system matrix for the region is given by

$$A = \begin{bmatrix} A_{LC} & -c \\ K_P E & 0 \end{bmatrix} \quad (5.43)$$

The augmented state variables within each area to be used throughout this chapter are defined as

$$x \triangleq \begin{bmatrix} x_{LC} \\ P_G \end{bmatrix} \quad (5.44)$$

instead of the traditionally used

$$x_{old} = \begin{bmatrix} x_{LC} \\ \delta_G \end{bmatrix} \quad (5.45)$$

The state coordinate transformation at each machine level is the mapping from the variables defined in (5.45) into (5.44). With these new state variables the general structure of an arbitrarily interconnected system is given in (5.42). At this point we could directly use the model formulation to re-visit the question regarding information obtained from the static load flow equations in analyzing small-signal stability. This was recently studied in [86]. It follows in a straightforward way that the small-signal stability properties of an isolated system ($F_e = 0$) are dependent only on the properties of the static network constraints, i.e., on the properties of matrix $K_P E$, when the local system matrices A_{LC} of all components are stable. This observation shows that as long as the local dynamics at each component level are stable, the small-signal stability information obtainable from matrix A is equivalent to the information obtainable from the static load flow equations. This claim is independent of the complexity of particular components.

From a structural point of view, the choice of electric real power outputs of the generators P_G as the state variables makes sense because they can be directly interpreted in terms of generators' interactions with the transmission system. Note that this model is not a simple generalization of the swing equations often used for transient stability analysis. The state variable used here is P_G instead of the state δ_G commonly employed in the swing equations. The significance of the new choice of state variables and their relation with the traditional ones is studied next.

5.3 Analysis

In this section, a detailed analysis of the real power/frequency dynamics is given, in much the same spirit as in Chapter 4. We first exploit the structural singularity of the real power/frequency dynamics.

5.3.1 Network properties

The transmission network has distinct properties that directly contribute to the inter-area oscillations and other important dynamic features. The decoupled Jacobian matrix J is defined as

$$J \triangleq \frac{\partial P^N}{\partial \delta} = \begin{bmatrix} J_{GG} & J_{GL} \\ J_{LG} & J_{LL} \end{bmatrix} \tag{5.46}$$

with submatrices J_{ij}, $i,j = G, L$ shown in (5.27) and (5.28). Define further

$$K_P^b \triangleq K_P|_{\text{No electrical losses}} \tag{5.47}$$

when the network is (real power) lossless. With these definitions, some important network properties are listed in the following proposition.

- **Proposition 5.1 (Network Properties):** For any (portion of a) network, the following is true:
    - $J\mathbf{1} = 0$
    - $K_P\mathbf{1} = 0$
    - $C_\omega \mathbf{1} = \mathbf{1}$
    - $\mathbf{1}^T K_P^b = 0$

    where $\mathbf{1}$ is the column vector with all 1's such that all operations are meaningful.

All the above properties are a direct consequence of the fact that the row sum of the incidence matrix is zero [87]. Note that $K_P\mathbf{1} = 0$ simply implies that, for any network, K_P is singular with $\mathbf{1}$ as the right eigenvector corresponding to its zero eigenvalue. The last property states further that, for a lossless network, matrix K_P^b has $\mathbf{1}^T$ also as its left eigenvector corresponding to the zero eigenvalue.

5.3.2 Structural singularity

The singularity of matrix K_P for any network leads to a fundamental characterization of regional system dynamics. For this, we have

- **Proposition 5.2 (Structural Singularity):** For any (portion of a) network, lossy or lossless, any generator type, the system matrix A defined in (5.43) is always singular.

This is a direct result of the singularity of K_P. Let the left eigenvector of K_P corresponding to its zero eigenvalue be l^T, i.e.,

$$l^T K_P = 0 \tag{5.48}$$

Consider the row vector

$$L \triangleq [\ 0 \ \ l^T\ ] \tag{5.49}$$

with the same number of zeros as the dimension of x_{LC}. We can check that

$$LA = [\ 0 \ \ l^T\ ] \begin{bmatrix} A_{LC} & -c \\ K_P E & 0 \end{bmatrix} = [\ l^T K_P E \ \ 0\ ] = 0 \tag{5.50}$$

i.e., matrix A is also singular with the vector L given in (5.49) as its left eigenvector corresponding to the zero eigenvalue. Details of the systemwide effect of this structural singularity are discussed in the next section.

5.3.3 Inter-area dynamics

Interaction variables were defined in Definition 4.1. It is emphasized that with the structure-based modeling approach, no weak interconnection assumption is required. To derive the inter-area dynamic model, we recall that the state space model of a single region is given by

$$\begin{bmatrix} \dot{x}_{LC} \\ \dot{P}_G \end{bmatrix} = \begin{bmatrix} A_{LC} & -c \\ K_P E & 0 \end{bmatrix} \begin{bmatrix} x_{LC} \\ P_G \end{bmatrix} - \begin{bmatrix} 0 \\ \dot{F}_e \end{bmatrix} + \begin{bmatrix} 0 \\ D_P \end{bmatrix} \dot{P}_L \tag{5.51}$$

with the system matrix

$$A = \begin{bmatrix} A_{LC} & -c \\ K_P E & 0 \end{bmatrix} \tag{5.52}$$

Let the inter-area variable be $z = Tx$. It has been shown that

$$T = L = \begin{bmatrix} 0 & l^T \end{bmatrix} \tag{5.53}$$

where $l^T K_P = 0$. The inter-area variable becomes

$$z = Tx = l^T P_G \tag{5.54}$$

This inter-area variable has a clear physical meaning: it is the combination of the real power outputs of all generators in the region according to the left eigenvector of the K_P matrix. This particular combination remains constant if the interactions are removed and there is no load variation. As tie-line flows or loads vary, the inter-area variable will vary with time.

The dynamic model for the inter-area variable is of the form

$$\dot{z} = -T \begin{bmatrix} 0 \\ \dot{F}_e \end{bmatrix} + T \begin{bmatrix} 0 \\ D_P \end{bmatrix} \dot{P}_L \tag{5.55}$$

or simply

$$\dot{z} = -l^T (\dot{F}_e - D_P \dot{P}_L) \tag{5.56}$$

Clearly it is seen that the inter-area variable z varies because of the tie-line flows, for a constant power load. Eq. (5.56) defines exactly the relation between these two. It is also seen that the inter-area dynamics represent the regional net real power exchanges. These kinds of interactions among subsystems are referred to as *inter-area power interactions*.

5.3.4 Computation of inter-area variables

Since the inter-area variables are defined for each individual separate region, the computation of inter-area variables can be done by each region separately. For each region, the computation involves only the calculation of the left eigenvector L of matrix A corresponding to its zero eigenvalues, as specified by

$$LA = 0 \tag{5.57}$$

In general, it is desired to have stable local dynamics, i.e., the governor control design is such that each matrix A_{LC} is of full rank. This assumption is normally met, since local controls are typically designed such that the local dynamics are stable. Under this condition, the singularity of matrix A is directly caused by the singularity of matrix K_P. In this case, we can take L as

$$L = \begin{bmatrix} 0 & l^T \end{bmatrix} \tag{5.58}$$

where l^T is the left eigenvector of matrix K_P corresponding to its zero eigenvalue, i.e.,

$$l^T K_P = 0 \tag{5.59}$$

Let us discuss the calculation of l^T. It is emphasized that (5.59) can be solved by a simple Gaussian elimination method. But a structural approach for the solution seems more physically meaningful. It was shown in Proposition 5.1. that, for a lossless network, we have a surprisingly simple but meaningful solution $l^T = \mathbf{1}^T$. This particular structure is by no means coincidental, it represents a fundamental requirement for the network—the power balance.

For a (real power) lossy network, since the losses are in general very small, we can use a perturbation method to obtain an approximate solution based on the lossless solution $l^T = \mathbf{1}^T$. The approximate solution is expected to be quite accurate if the losses are not unreasonably high. A numerical example will show later that the approximation gives a simple but quite satisfactory answer. Again, this approach is a (more physically meaningful) alternative to the numerical Gaussian method. Let us write l as

$$l = l^b - l^g = \mathbf{1} - l^g \tag{5.60}$$

where $l^b = \mathbf{1}$ corresponds to the eigenvector of the lossless K_P matrix, and l^g is added to account for losses. When the losses are small, the term l^g is expected to be small. Condition (5.59) becomes

$$l^T K_P = (l^b - l^g)^T (K_P^b + K_P^g) = 0 \tag{5.61}$$

Recognizing from Proposition 5.1. that $l^b K_P^b = 0$, and neglecting the higher-order term $l^{gT} K_P^g$ due to small losses, one obtains

$$l^{gT} K_P^b = l^{bT} K_P^g \tag{5.62}$$

Furthermore, without loss of generality, one could assume the form for l^g to be

$$l^{gT} = \begin{bmatrix} 0 & l_2^g & \cdots & l_m^g \end{bmatrix} \tag{5.63}$$

since the transformation l^T is unique only up to a constant scalar, and one can always choose the first element of l^T to be the first element of l^{bT}, and therefore the first element of l^{gT} is always 0.

Let us denote

$$l_-^{gT} \triangleq \begin{bmatrix} l_2^g & \cdots & l_m^g \end{bmatrix} \tag{5.64}$$

Then (5.62) can be written as

$$\begin{bmatrix} 0 & l_-^{gT} \end{bmatrix} K_P^b = l^{bT} K_P^g \tag{5.65}$$

Since the first row of K_P^b is multiplied by zero, it can be eliminated. Define matrix

$$E_1 \triangleq I \Big|_{\text{Without 1st Column}} \tag{5.66}$$

Then, for any matrix M,

$$ME_1 = M \big|_{\text{Without 1st Column}} \tag{5.67}$$
$$E_1^T M = M \big|_{\text{Without 1st Row}} \tag{5.68}$$

With this notation, (5.65) now becomes

$$l_-^{gT} E_1^T K_P^b = l^{bT} K_P^g \tag{5.69}$$

Further, let us split K_P^b and K_P^g as follows:

$$K_P^b = \begin{bmatrix} K_{P1}^b & \vdots & K_{P-}^b \end{bmatrix} = \begin{bmatrix} K_{P1}^b & \vdots & K_P^b E_1 \end{bmatrix} \tag{5.70}$$

$$K_P^g = \begin{bmatrix} K_{P1}^g & \vdots & K_{P-}^g \end{bmatrix} = \begin{bmatrix} K_{P1}^g & \vdots & K_P^g E_1 \end{bmatrix} \tag{5.71}$$

where K_{P1}^b and K_{P1}^g represent the first column of K_P^b and K_P^g. Since both K_P^b and K_P^g satisfy

$$K_P^b \mathbf{1} = 0 \text{ and } K_P^g \mathbf{1} = 0 \tag{5.72}$$

from Proposition 5.1., we have

$$K_{P1}^b = -K_P^b E_1 \mathbf{1} \text{ and } K_{P1}^g = -K_P^g E_1 \mathbf{1} \tag{5.73}$$

Thus,

$$K_P^b = \begin{bmatrix} -K_P^b E_1 \mathbf{1} & \vdots & K_P^b E_1 \end{bmatrix} \tag{5.74}$$

$$K_P^g = \begin{bmatrix} -K_P^g E_1 \mathbf{1} & \vdots & K_P^g E_1 \end{bmatrix} \tag{5.75}$$

Eq. (5.69) becomes

$$l_-^{gT} E_1^T \begin{bmatrix} -K_P^b E_1 \mathbf{1} & \vdots & K_P^b E_1 \end{bmatrix} = l^{bT} \begin{bmatrix} -K_P^g E_1 \mathbf{1} & \vdots & K_P^g E_1 \end{bmatrix} \tag{5.76}$$

or

$$l_-^{gT} E_1^T K_P^b E_1 \mathbf{1} = l^{bT} K_P^g E_1 \mathbf{1} \tag{5.77}$$

$$l_-^{gT} E_1^T K_P^b E_1 = l^{bT} K_P^g E_1 \tag{5.78}$$

Eq. (5.77) will be automatically satisfied if (5.78) is satisfied. This redundancy is always the case because of the singularity of K_P.

Let us now consider (5.78) only. Under normal operating conditions, matrix K_P^b is rank-deficient only by 1. In this case, matrix $E_1^T K_P^b E_1$, which is the matrix K_P^b with first row and first column eliminated, has full rank. Therefore, one can solve for l_-^{gT} from (5.78) as

$$l_-^{gT} = l^{bT} K_P^g E_1 (E_1^T K_P^b E_1)^{-1} \tag{5.79}$$

Using the fact that $l^b = \mathbf{1}$, one derives

$$l^{gT} = \begin{bmatrix} 0 & \vdots & \mathbf{1}^T K_P^g E_1 (E_1^T K_P^b E_1)^{-1} \end{bmatrix} \tag{5.80}$$

This is the desired expression for the perturbation to the left eigenvector of K_P corresponding to its zero eigenvalue. It is valid for small real power losses.

It is interesting to recognize that when matrix K_P has rank lower than $(m-1)$, there exists a possibility of having more than one interaction variable per area. These additional interaction variables are simply caused by the nonexistence of any solution to the static network constraints (5.21) and are independent of the relative inertia and damping coefficient values of generators and of the type of governor controls. Although a question remains if such operating points would be feasible, a near loss of rank would be of definite practical interest. For example, one scenario of additional loss of rank in the matrix K_P would be when the system operates at unusually high real power transfers. The inter-area oscillations have been recognized in the context of this operating mode [80].

5.3.5 Interpretation of inter-area variables

Let us now discuss the physical meaning of the inter-area variables defined in Definition 4.1. Since inter-area variables are defined when interconnections are removed, i.e., they are defined separately for each region, we focus now on a single region. First, for the (real power) lossless case, it has been shown that $l^T = \mathbf{1}^T$. The inter-area variable can, from (5.54), be further written as

$$z = l^T P_G = \mathbf{1}^T P_G = \sum_{j=1}^{m} P_e^j \tag{5.81}$$

i.e., the inter-area variable is the sum of the real power outputs of all the generators in this region, or the total generation of the region. Under the conditions specified in Definition 4.1, i.e., when there are no tie-line flow exchanges with neighboring regions, and no load variations, the total generation of the region must be constant, equaling the constant load demand. It is clear that the definition of the inter-area variables captures this fundamental property of the network.

- **Proposition 5.3 (ACE as a particular case of an inter-area variable):** The inter-area variable z defined in (5.81) is equivalent to the ACE of the same area used at present in AGC, under the following assumptions:

  1. The system frequency is nominal, and
  2. The system losses are neglected.

Proof: The ACE defined in (3.34) when combined with the relation (5.24).

For the case of a lossy network, part of the generation must be absorbed by the losses in the network, and the other part of the generation must be balanced with the loads. Therefore, the total generation of the region minus the power absorbed by the losses must be a constant under the conditions in Definition 4.1. This portion of the power absorbed by the losses is precisely described by the correction term $l^{gT}P_G$. In fact it can be shown that the term $l^{gT}P_G$ is exactly the total real power loss of the region. Thus we see here again the requirement of power balance for the network. This power balance requirement is the fundamental property of the network. It will have a systemwide effect on higher-level dynamics also.

5.3.6 Comparisons with conventional models

The choice of real power outputs P_G of generators as state variables greatly facilitates the study of inter-area dynamics. For comparison, we examine conventional models, which employ generator phase angles as state variables. The local dynamic model, under the condition that updating the reference values is inactive, has been obtained as

$$\dot{x}_{LC} = A_{LC}x_{LC} - cP_G \tag{5.82}$$

where the generator power P_G is replaced by the network coupling relation in (5.36):

$$P_G = K_P\delta_G + D_P P_L - F_e \tag{5.83}$$

Together, the global linearized model of an area takes the form

$$\dot{x}_{LC} = A_{LC}x_{LC} - c(K_P\delta_G + D_P P_L - F_e) \tag{5.84}$$

Chapter 5   99

and the trivial relation

$$\dot{\delta}_G = \omega_G = Ex_{LC} \tag{5.85}$$

is used. The most often used model with conventional state variables is the so-called swing dynamic model, i.e., there is no governor control, and the mechanical power applied to the generator shaft is constant. The primary dynamics become

$$M\ddot{\delta}_G + D\dot{\delta}_G + K_P\delta_G + D_P P_L - F_e = 0 \tag{5.86}$$

with

$$M = \begin{bmatrix} M^1 & & \\ & \ddots & \\ & & M^m \end{bmatrix} \quad \text{and} \quad D = \begin{bmatrix} D^1 & & \\ & \ddots & \\ & & D^m \end{bmatrix} \tag{5.87}$$

being the inertia and damping matrices. This model loses the clear structural properties of system matrix A, as stated in Proposition 5.2. Also, it is not easily generalized to more complicated cases.

The new state variables P_G can be viewed as a linear transformation from the conventional state variables δ_G as specified in (5.83)

$$P_G = K_P \delta_G + D_P P_L - F_e \tag{5.88}$$

or simply

$$P_G = K_P \delta_G \tag{5.89}$$

if one neglects the inputs to the system. This is not a conventional transformation in the sense that the transformation matrix K_P is always singular for any network, as stated in Proposition 5.1. This singular transformation proves to be very efficient for the basic understanding of inter-area oscillations in power systems.

Let us first briefly present the existing results in the literature. For the conventional model (5.86), storage variables have been proposed in [69]. These storage variables are defined as

$$\xi = M\omega_G + D\delta_G \tag{5.90}$$

Note that the defined storage variables have the dimension of the number of generators, not just a single scalar for an area. The question is how the generalized definition of the interaction variables is related to this widely accepted approach to modeling inter-area variables. To answer this, we first interpret the previous work from the viewpoint of Definition 4.1.

If we perform the calculation of the inter-area variables for the model (5.86) in the old state variables, it can be shown that the resulting area variable is

$$z_{old} = l^T \xi = l^T(M\omega_G + D\delta_G) \tag{5.91}$$

where $l^T K_P = 0$, the same left eigenvector of P_G corresponding to the zero eigenvalue. In other words, the scalar inter-area variable defined here when applied to the conventional model is the same linear combination of the storage variables as the inter-area variable

$$z = l^T P_G \tag{5.92}$$

in the new state space proposed here, with the left eigenvector l^T of P_G. The relation between the two inter-area variables is derived from (5.86) (with $P_L = 0$ and $F_e = 0$) as

$$z = -\dot{z}_{old} \tag{5.93}$$

since

$$P_G = K_P \delta_G = -(M\dot{\omega}_G + D\omega_G) = -\dot{\xi} \tag{5.94}$$

In the context of mechanical variables, the inter-area variable z can be explained as the force, and the inter-area variable z_{old} can be viewed as the momentum modulo mechanical losses.

There are a few drawbacks with the conventional model. Conceptually, first, the inter-area variable $z_{old} = l^T(M\omega_G + D\delta_G)$ misleadingly indicates that generator local dynamics directly participate in the inter-area mode, since all quantities involved are mechanical ones constituting the generator local dynamics. Although these local dynamics quantities show up in the inter-area dynamic equation, the particular expression $(M\omega_G + D\delta_G)$ actually implies the electrical power balance.

Second, structurally, the form $z_{old} = l^T(M\omega_G + D\delta_G)$ is valid only for linear damping. When the damping takes a nonlinear form, the expression is completely invalid. But the inter-area variable $z = l^T P_G$ still holds. This is because the inter-area variable reflects the fundamental requirement of the power (force) balance for an isolated system without any disturbances. The expression $(M\omega_G + D\delta_G)$ is useful just because it is another form of the power (force) balance, for

$$(M\omega_G + D\delta_G) = -\int P_G(\tau)d\tau \tag{5.95}$$

through the dynamic equation. For different forms of damping, expression $(M\omega_G + D\delta_G)$ becomes meaningless, while the basic power (force) balance

Chapter 5

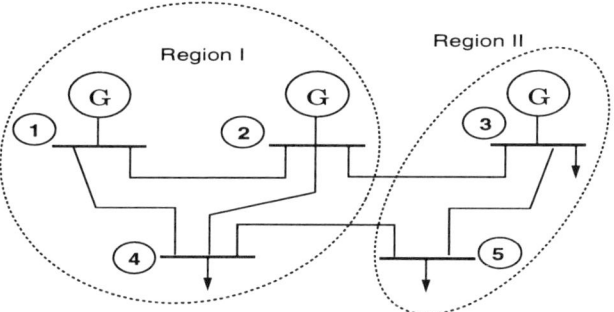

Figure 5.3: An interconnected 5-bus power system

will still hold. This clearly reveals that the choice of the power outputs P_G as state variables captures the fundamental properties of the system.

Third, from the point of view of control design to suppress inter-area oscillations, the inter-area variable $z_{old} = l^T(M\omega_G + D\delta_G)$ involves the measurements of m angles δ_G and m frequencies ω_G, and the inertia matrix M and the damping matrix D. It is very hard to get accurate numerical values for these parameters. In the new state space, the inter-area variable $z = l^T P_G$ only involves the measurements of m generator power outputs, which are measured anyway. No parameters regarding the local dynamics are needed.

5.3.7 An example

In this section, we study a small example power system to illustrate the theoretical concepts developed in this chapter. The example is a 5-bus power system, shown in Figure 5.3.

This example will be used throughout this chapter consistently to show the structural decompositions and control designs. The system is composed of two regions. Region I includes generator buses #1 and #2 and load bus #4. Region II includes generator #3 and load #5. There are two tie lines connecting generator #4 to #3, and load #4 to #5. The data used in the simulation are listed in Table 5.1.

All simulations are done using the software MATLAB. The K_P matrix for each region is calculated as

$$K_P^I = \begin{bmatrix} 14.8977 & -14.8977 \\ -14.9446 & 14.9446 \end{bmatrix} \quad \text{and} \quad K_P^{II} = 0 \tag{5.96}$$

Clearly, each row in the K_P matrix sums to zero, i.e.,

$$K_P^I \begin{bmatrix} 1 \\ 1 \end{bmatrix} = 0 \quad \text{and} \quad K_P^{II} \mathbf{1} = 0 \tag{5.97}$$

Table 5.1: Per-unit data for the 5-bus example

	\multicolumn{7}{c}{G-T-G Set Parameters}						
	M	D	e_T	T_u	K_t	r	T_g
#1	10	5	1696	.2	10744	19	.25
#2	5	4	1696	.2	10744	19	.25
#3	3	4	1696	.2	10744	19	.25
	\multicolumn{7}{c}{Line Parameters}						
	#1 – #2	#1 – #4	#2 – #4	#3 – #5	#2 – #3	#4 – #5	
b	10	10	10	10	.5	.5	
g	1	1	1	1	.05	.05	
	\multicolumn{7}{c}{Load Flow Data}						
	#1	#2	#3	#4	#5		
V	1	1	1	.9802	.9502		
δ	0	0.0157	-0.191	-0.0575	-0.3013		

Table 5.2: Closed-loop eigenvalues—lower damping

	\multicolumn{2}{c}{Region I}	\multicolumn{2}{c}{Region II}		
Disconnected System	$-0.3487 \pm 2.0825j$		-0.6025	0	-1.3333	0
Connected System	$-0.3509 \pm 2.100j$		$-0.4201 \pm 0.2503j$		-1.5523	0

It is easy to check that the left eigenvector of K_P corresponding to the zero eigenvalue is given by

$$l^{I\,T} = \begin{bmatrix} 1 & 0.9969 \end{bmatrix} \quad \text{and} \quad l^{II\,T} = 1 \tag{5.98}$$

For the purposes of illustration, let us consider only the swing dynamics. In this case, the closed-loop eigenvalues for the disconnected and connected systems are computed and shown in Table 5.2.

From Table 5.2, it is seen that each disconnected region has a zero eigenvalue. One zero eigenvalue remains at zero when the regions are connected, while two eigenvalues -0.6025 and 0 move together to the complex eigenvalues $-0.4201 \pm 0.2503j$. These two additional oscillatory eigenvalues correspond to the inter-area oscillations. The frequency of the inter-area oscillations $\omega = 0.2503$ is much smaller than the other frequency $\Omega = 2.100$ to yield the so-called slow intra-area oscillations.

Figure 5.4 shows the response of the system. Obviously, both fast intra-area secondary- and slow inter-area oscillations are seen. As expected, the inter-area variable z^I captures the slow inter-area oscillations.

5.4 Model derivations

5.4.1 Quasi-static model

In this section, a simple discrete time dynamic model on the secondary time scale is derived. The purpose of regulation at the regional secondary level is to update the frequency reference value for each participating G-T-G set at discrete time instant kT_s so that steady-state frequency errors are eliminated. The discrete-time actions of updating frequency reference values result in a discrete event process (DEP) in frequencies on the secondary level.

Let us start with the closed-loop primary dynamic model (5.42):

$$\begin{bmatrix} \dot{x}_{LC} \\ \dot{P}_G \end{bmatrix} = \begin{bmatrix} A_{LC} & -c \\ K_P E & 0 \end{bmatrix} \begin{bmatrix} x_{LC} \\ P_G \end{bmatrix} + \begin{bmatrix} b \\ 0 \end{bmatrix} \omega_G^{ref}[k] - \begin{bmatrix} 0 \\ \dot{F}_e \end{bmatrix}$$
$$+ \begin{bmatrix} 0 \\ D_P \dot{P}_L \end{bmatrix} \tag{5.99}$$

Assume that the governor controls are designed such that the closed-loop transient dynamics are fast relative to the updating of reference values. Under this assumption one can write $\dot{x} = 0$, at kT_s, $k = 0, 1, \ldots$, i.e., the system settles to a steady state at these discrete time instants kT_s. Let us first consider the local dynamics of each G-T-G set, derived in (5.19):

$$\dot{x}_{LC} = A_{LC} x_{LC} + b\omega_G^{ref}[k] - cP_G \tag{5.100}$$

The assumption of fast transient dynamics yields

$$A_{LC} x_{LC}[k] = -b\omega_G^{ref}[k] + cP_G[k] \tag{5.101}$$

or

$$x_{LC}[k] = -A_{LC}^{-1} b\omega_{ref}[k] + A_{LC}^{-1} cP_G[k] \tag{5.102}$$

since A_{LC}, given in (5.12), is invertible. One can further calculate

$$A_{LC}^{-1} = \frac{1}{\Delta} \begin{bmatrix} -rM & -rT_u & -(e_T + K_t)T_g \\ K_t M & -(rD + e_T)T_u & -DK_t T_g \\ M & T_u & -DT_g \end{bmatrix} \tag{5.103}$$

with $\eta = rD + e_T + K_t$. Substituting this back into (5.102) simply gives

$$\begin{bmatrix} \omega[k] \\ P_m[k] \\ a[k] \end{bmatrix} = \frac{1}{\eta} \begin{bmatrix} e_T + K_t \\ DK_t \\ D \end{bmatrix} \omega^{ref}[k] + \frac{1}{\eta} \begin{bmatrix} -r \\ K_t \\ 1 \end{bmatrix} P_G[k] \tag{5.104}$$

Since we are interested in the frequency variations only on the secondary level, let us consider only the first row,

$$\omega[k] = \frac{e_T + K_t}{\eta} \omega^{ref}[k] - \frac{r}{\eta} P_G[k] \tag{5.105}$$

or

$$\omega[k] = (1 - \frac{r}{\eta} D)\omega^{ref}[k] - \frac{r}{\eta} P_G[k] \tag{5.106}$$

Define the *droop constant* of any G-T-G set as

$$\sigma \triangleq - \frac{\partial \omega[k]}{\partial P_G[k]} \bigg|_{\omega^{ref}[k]=0} \qquad (5.107)$$

The physical meaning of the droop constant is that it represents the sensitivity of the steady-state frequency deviation to the deviation of steady-state real power output of the G-T-G set, when the frequency setting is kept at a constant or the secondary-level control is inactive. A small droop constant indicates that real power output variations have a small effect on the steady-state frequency variations. A flat droop characteristic, i.e., $\sigma = 0$, implies that the steady-state frequency always reaches the reference value, no matter how the real power output varies. Clearly, an integral control must be involved in the primary controller in this case, so that the steady-state error of the primary local control vanishes.

With this general definition, one simply obtains the droop constant for the G-T-G set discussed here as

$$\sigma = \frac{r}{\eta} = \frac{r}{rD + e_T + K_t} \qquad (5.108)$$

It then follows from (5.11) that the steady-state transfer function of the G-T-G set for specified $P_G[k]$ is

$$\omega[k] = (1 - \sigma D)\omega^{ref}[k] - \sigma P_G[k] \qquad (5.109)$$

where D is the generator damping constant.

This is the quasi-static relation among the frequency deviation, reference value, and real power output variation of each G-T-G set. This relation is completely decentralized, in the sense that no coupling between different G-T-G sets occurs except through the local output variables P_G. This complete decentralization is possible only if the real power output of each G-T-G set is chosen as a state variable, as is done in this chapter. Real power output variations of all G-T-G sets in the network will be coupled, and the coupling is exactly the network relation discussed before.

Let us derive the secondary-level relation for a region consisting of m such G-T-G sets. Since (5.109) is completely decentralized and true for each G-T-G set, the secondary-level relation for the entire region is obtained by simply gathering the relations (5.109) for each G-T-G set. To do this, we define the generator frequency vector and power output vectors as

$$\omega_G \triangleq \begin{bmatrix} \omega_G^1 \\ \vdots \\ \omega_G^m \end{bmatrix} \qquad (5.110)$$

and the diagonal droop matrix and damping matrix as

$$\sigma \triangleq \begin{bmatrix} \sigma_1 & & \\ & \ddots & \\ & & \sigma_m \end{bmatrix} \quad \text{and} \quad D \triangleq \begin{bmatrix} D_1 & & \\ & \ddots & \\ & & D_m \end{bmatrix} \tag{5.111}$$

Then the decentralized quasi-static model for all m G-T-G sets can be obtained as

$$\omega_G[k] = (I - \sigma D)\omega_G^{ref}[k] - \sigma P_G[k] \tag{5.112}$$

where I is the $m \times m$ identity matrix.

Vector $P_G[k]$ is coupled to all generator angles by the following equation derived in (5.83):

$$P_G[k] = K_P \delta_G[k] - F_e[k] + D_P P_L[k] \tag{5.113}$$

Eq. (5.112) is combined with (5.113) to yield

$$\omega_G[k] = (I - \sigma D)\omega_G^{ref}[k] - \sigma(K_P \delta_G[k] - F_e[k] + D_P P_L[k]) \tag{5.114}$$

Writing (5.114) at two successive load frequency control (LFC) sampling instants kT_s and $(k+1)T_s$, one obtains

$$\begin{aligned} \omega_G[k+1] - \omega_G[k] &= (I - \sigma D)(\omega_G^{ref}[k+1] - \omega_G^{ref}[k]) \\ &\quad - \sigma K_P(\delta_G[k+1] - \delta_G[k]) + \sigma(F_e[k+1] - F_e[k]) - \\ &\quad \sigma D_P(P_L[k+1] - P_L[k]) \end{aligned} \tag{5.115}$$

Since

$$\delta_G[k+1] - \delta_G[k] \approx T_s \omega_G[k] \tag{5.116}$$

model (5.115) expressed in terms of $\omega_G[k]$ only takes the form

$$\begin{aligned} \omega_G[k+1] &= (I - \sigma K_P T_s)\omega_G[k] + (I - \sigma D)(\omega_G^{ref}[k+1] - \omega_G^{ref}[k]) \\ &\quad + \sigma(F_e[k+1] - F_e[k]) - \sigma D_P(P[k+1] - P[k]) \end{aligned} \tag{5.117}$$

Model (5.117) is defined in terms of system variables at discrete times kT_s, $k = 0, 1, \ldots$ only. The general theory of control design for such systems [90] is applicable to this model.

The discrete-time corrective signal $(\omega_G^{ref}[k+1] - \omega_G^{ref}[k])$ is the control action for the secondary level. To allow for generating units not participating in secondary-level control, let us separate the participating and nonparticipating generators as

$$\omega_G = \begin{bmatrix} \omega_s \\ \omega_n \end{bmatrix} \tag{5.118}$$

where ω_s and ω_n represent the participating and nonparticipating generators, respectively. The term $(\omega_G^{ref}[k+1] - \omega_G^{ref}[k])$ can now be written as

$$\omega_G^{ref}[k+1] - \omega_G^{ref}[k] = \begin{bmatrix} \omega_s^{ref}[k+1] - \omega_s^{ref}[k] \\ \omega_n^{ref}[k+1] - \omega_n^{ref}[k] \end{bmatrix} \quad (5.119)$$

It is clear that for the nonparticipating generators

$$\omega_n^{ref}[k+1] - \omega_n^{ref}[k] \equiv 0, \quad \forall\, k \quad (5.120)$$

Let us define the actual secondary-level LFC signal as

$$u_s[k] \triangleq \omega_s^{ref}[k+1] - \omega_s^{ref}[k] \quad (5.121)$$

From these definitions, we can rewrite the control term as

$$\omega_G^{ref}[k+1] - \omega_G^{ref}[k] = \begin{bmatrix} u_s[k] \\ 0 \end{bmatrix} = \begin{bmatrix} I \\ 0 \end{bmatrix} u_s[k] \triangleq B_s u_s[k] \quad (5.122)$$

where I is the $p \times p$ identity matrix, with p being the number of participating generators.

Let us further define the net tie-line flow effect as

$$f[k] = F_e[k+1] - F_e[k] \quad (5.123)$$

and also define the disturbance at the secondary level as

$$d_s[k] = P_L[k+1] - P_L[k] \quad (5.124)$$

With these definitions, we derive the discrete-time dynamics of the generator frequencies at the secondary level as

$$\omega_G[k+1] = (I - \sigma K_P T_s)\omega_G[k] + (I - \sigma D)B_s u_s[k] + \sigma f[k] - \sigma D_P d_s[k] \quad (5.125)$$

This is the secondary level dynamic model for all generator frequencies in terms of the frequency setting changes and tie-line flow changes.

Load frequencies ω_L are expressed in terms of the generator frequencies ω_G by (5.32):

$$\omega_L = C_\omega \omega_G + J_{LL}^{-1}(\dot{F}_L - \dot{P}_L) \quad (5.126)$$

which leads to, with all time derivatives vanishing at discrete times kT_s,

$$\omega_L[k] = C_\omega \omega_G[k] \quad (5.127)$$

This relation determines changes in load frequencies in terms of changes in generator frequencies.

The output variables for control at the secondary level include part or all of the generators and possibly some loads. Inclusion of load frequencies allows for demand-side management of secondary-level frequency regulations. Let us define the output variables as

$$\omega_o[k] = C_1 \omega_G[k] + C_2 \omega_L[k] \tag{5.128}$$

where C_1 and C_2 are matrices with 0's and 1's to pick up the desired output variables. Using (5.127) to express $\omega_L[k]$ in terms of the state variables $\omega_G[k]$, we obtain the output equation of secondary-level frequency control as

$$\omega_o[k] = C_s \omega_G[k] \tag{5.129}$$

where the output matrix is simply $C_s = C_1 + C_2 C_\omega$.

Eqs. (5.125) and (5.129) constitute the simple discrete-time dynamic model for variables of interest at the secondary level. This simple model forms the basis for frequency regulation at the secondary level.

We now use a numerical example to confirm that the derived simple quasi-static models agree with the detailed transient models evaluated on the secondary time scale T_s. In the simulation, secondary-level control is not activated, i.e., $u_s[k] = 0$, $\forall k$. The entire system is considered as a single area. Because this region is isolated, there is no tie-line flow. We compare the time domain responses of the system under a given load variation, obtained using the full complex primary-level models and the much simpler secondary-level models. All initial conditions are set to zero. Figure 5.5 shows the linearly varying load at bus #4 and the responses of three generator frequencies to this load variation with the primary and secondary models. The following is observed from the numerical simulations:

- Frequencies of generators #1 and #2 are almost equal. This is because the two lines connecting bus #4 to generators #1 and #2 are the same. Generator #3 is nearly unperturbed by the load disturbance occurring at bus #4 because the two tie lines are weak.

- Time responses of generator frequencies roughly follow inversely the load fluctuations, i.e., load increases result in frequency decreases, and vice versa.

- Time domain responses of the system obtained using complete model (5.42) are indistinguishable from responses obtained using much simpler models (5.125) and (5.129). However, the calculations for the quasi-static models are appreciably simpler and easier than the calculations for the original detailed models.

5.4.2 Generator power model

A model with generator output power, $\underline{P}_G$, as states can be derived using a similar approach [75]. This model is

$$\begin{aligned}
\underline{P}_G[k+1] &= (I - K_P\sigma T_s)\underline{P}_G[k] + \\
&\quad K_P\left(I - \sigma\bar{D}\right)T_s\underline{\omega}_G^{ref}[k] - \sigma\left(f[k] - D_p\underline{d}[k]\right)
\end{aligned} \qquad (5.130)$$

This model is useful for tracking the evolution of generator power in response to governor reference settings, $\underline{\omega}_G^{ref}[k]$.

An important qualitative difference between the model given in terms of frequencies $\omega_G[k]$ as state variables (5.125) and the model (5.130) in which $P_G[k]$ are the state variables is in the following. Model (5.125) is solely control-driven in the sense of Definition 4.1 since when there are no disturbances in load and tie-line flows $u_s[k] \equiv 0$ leads to $\omega_G[k+1] = \omega_G[k]$. The secondary-level dynamic model (5.130), expressed in terms of $P_G[k]$, is not a control-driven model because of the droop characteristics of the generator primary control [4]. An in-depth analysis shows that this lack of full controllability in the P_G state space is a direct consequence of the structural singularity defined in Section 5.3.2. As a result, state variables $P_G[k]$ can still evolve with time even when the system is subject to no tie-line flows and disturbances and $u_s[k] \equiv 0$, as a consequence of the generator dynamics.

One way to interpret the practical consequences of this mathematical property is to realize that the electric power system is inherently a floating system, i.e., that neither the primary frequency controllers nor present automatic generation control (AGC) are capable of retaining the absolute system frequency at its nominal value [32]. Strictly speaking, this is a direct consequence of the structural singularity of the power system model described in Chapter 4. The present mechanism used to get around this issue and maintain the absolute frequency at 60 Hz is by means of a time-correction mechanism at a prespecified point on a very large system. Whenever the cumulative effect of frequency deviations exceeds a prespecified amount, the generation is adapted to correct for these deviations. This is a peculiar phenomenon and could cause much trouble, if not understood clearly, particularly in a changing industry. Rules must be established concerning the responsibility and value of maintaining absolute frequency close to its nominal value.

A centralized model for tertiary-level real power tie-line flow regulation can be found in [74, 77, 147]. Note that this is a particular case of model (4.51)–(4.52) expressed in terms of the interaction variables defined in Chapter 4. The interaction variables are linear combinations of tie-line flows and as such have a straightforward physical interpretation.

5.5 Control design

It is shown in this section how both secondary-level models (5.125) and (5.130) can be used for meeting control performance objectives set by each horizontally structured subsystem under a changing industry. The tertiary-level model [74, 77] is best used at the interconnected system level, assuming the objectives at the lower levels are met.

For the present industry, this control design is useful for achieving the performance criteria described in Section 3.7.1 at each control area level. The tertiary-level model could be used to regulate tie-line flow exchanges, including inadvertent energy exchange, according to given prespecifications among the very large interconnected systems, such as large portions of the United States, while requiring certain fringe control at each subsystem level [147]. The coordination at the interconnected system level by means of tertiary level is minimal [147]. The proposed control design is also relevant for the nested hierarchy of a changing industry. It can be used to account for the contribution of independent power producers to real power/frequency regulation within a control area of an otherwise horizontally structured system. It could also be used for cost allocation at the secondary and tertiary levels of the system in response to various wheeling transactions, as well as for noncompliance [77, 78].

As mentioned earlier in Chapter 3, the economic transactions are simply viewed as positive or negative deviations in demand to which these control schemes respond. One should not lose sight of the need for more metering of various transactions and participants in the control schemes for the proposed unbundling to be feasible.

5.5.1 Regulating frequency at the secondary level

It is shown here that the fairly simple quasi-static dynamic model of frequency (5.125) is basic to developing decentralized secondary-level controllers. The proposed secondary-level control schemes provide system-theoretic support for AGC and help to answer many open questions, stated, for example, in [1, 6, 93].

The main task of regulation at the secondary level is to eliminate steady-state deviations of frequencies at critical locations from their scheduled values. Secondary-level control, often referred to as LFC, is implemented by updating the reference values of governor speed-changers of the G-T-G sets participating at this control level at discrete time instant kT_s, $k = 0, 1, \ldots$ In practice, LFC is often combined with generation scheduling needed to optimize total fuel cost for meeting a given load at each regional level. The simplest static optimization method for distributing total P_G among all generators according to their production costs to maintain the average frequency at each subsystem level is the economic dispatch method. This optimization task is generally not coordinated with LFC; this is one of the unsolved problems in the LFC area [2]. The mathematical formulation of optimal LFC proposed here overcomes this issue because the contribution of each generator to the average frequency is defined as depending on the electric properties of the transmission network.

Secondary-level control should keep operations at each regional level as autonomous as possible. In other words, secondary-level control should be designed so that deviations in mechanical outputs of the G-T-G sets are introduced to respond to the load deviations within the same region while suppressing unintentional changes in real power tie-line flows. This requirement is based on the widely accepted area control principle [1].

The secondary-level quasi-static discrete-time model for an administrative region within an interconnected system, derived in (5.125), and is repeated here:

$$\omega_G[k+1] = (I - \sigma K_P T_s)\omega_G[k] + (I - \sigma D)B_s u_s[k] + \sigma f[k] - \sigma D_P d_s[k] \quad (5.131)$$

The secondary-level control signal $u_s[k]$ is used to cancel the effect of the load and tie-line flow variations to eliminate steady-state frequency deviations. Since the number of tie lines is small, and tie-line flows are monitored in practice, a new control logic at the secondary level to stabilize frequencies is proposed as follows:

$$u_s[k] = G_s(\omega_o[k] - \omega_o^{set}[K]) + H_s f[k] \quad (5.132)$$

where $\omega_o^{set}[K]$ is the set value for the output frequencies defined in (5.129). The settings are changed on a longer time scale T_t by tertiary-level regulation to achieve best performance over the time horizon T_t. The gain matrices G_s and H_s are to be determined.

Under this control law, the closed-loop dynamics on the secondary level become

$$\omega_G[k+1] = (I - \sigma K_P T_s)\omega_G[k] + (I - \sigma D)B_s G_s(\omega_o[k] - \omega_o^{set}[K])$$

$$+[(I - \sigma D)B_s H_s + \sigma]f[k] - \sigma D_P d_s[k] \quad (5.133)$$

Using the output equation $\omega_o[k] = C_s \omega_G[k]$, one obtains

$$\omega_G[k+1] = A_s \omega_G[k] - (I - \sigma D)B_s G_s \omega_o^{set}[K] + \\ [(I - \sigma D)B_s H_s + \sigma]f[k] - \sigma D_P d_s[k] \quad (5.134)$$

where

$$A_s \triangleq (I - \sigma K_P T_s) + (I - \sigma D)B_s G_s C_s \quad (5.135)$$

is the closed-loop system matrix of secondary-level dynamics.

The purpose of the additional term $H_s f[k]$ in the control law is to cancel the effect of tie-line flows from the neighboring regions, so that each region has effectively decoupled dynamics. Full decoupling can be achieved if

$$(I - \sigma D)B_s H_s + \sigma = 0 \tag{5.136}$$

In general, both generator damping and the droop constants are very small, so that matrix $(I - \sigma D)$ is invertible. In this case, to derive a unique solution for H_s from (5.136), one must require that B_s be nonsingular. From the structure of B_s defined in (5.122), it is clear that nonsingularity of B_s is equivalent to all generators participating in secondary-level frequency control. In this case, $B_s = I$, and we can simply choose

$$H_s = -(I - \sigma D)^{-1} \sigma \tag{5.137}$$

to cancel the effect of neighboring regions. With this choice of H_s, the region under study looks as if it were disconnected from the rest of the system. The closed-loop dynamics of the region take on the form

$$\omega_G[k+1] = A_s \omega_G[k] - (I - \sigma D)B_s G_s \omega_o^{set}[K] - \sigma D_P d_s[k] \tag{5.138}$$

with no coupling among different regions. Unless all generators participate in secondary control, complete cancellation of tie-line flows in all generator frequencies is not possible. In this case, only partial cancellation for some state variables can be achieved.

The gain matrix G_s can be determined by specifying the desired closed-loop dynamics, as will be seen from the numerical example that follows, or by formulating an optimal control problem. A linear quadratic performance criterion for the region can be written as

$$J_s = \sum_{k=0}^{\infty} (\omega_o^T[k] Q \omega_o[k] + u_s^T[k] R u_s[k]) \tag{5.139}$$

for $Q = Q^T \geq 0$ and $R = R^T > 0$. Depending on the relative importance of the quality of frequency regulation and the fuel cost associated with specific G-T-G sets, the weighting matrices Q and R in the performance criterion will vary. The optimization with respect to the secondary controls $u_s[k]$ determines the optimal gain G_s. The generalized formulation here allows for including the frequencies of all generators and loads participating in LFC instead of the conventionally used criterion in terms of average frequency only [67, 81].

The performance criterion reflects specifications of the output variables at the regional level. It is sufficiently general to allow for specifying different frequency quality requirements at different individual generators or loads throughout the area. In light of the new regulatory constraints on operating power systems in the United States, this feature is taking on a new importance. The jointly owned units, non-utility-owned units, and large loads participating in demand-side management are potentially the points in the system that need such monitoring and that would belong to the set $\omega_o[k]$. Note that it is possible to relate generator frequency variations ω_G (as the output variable to which both primary and secondary controllers respond) to the load frequency variations ω_L as a relevant operating specification. The relation is the simple one derived in (5.127). This is not done at present, but with the formulation proposed here it is fairly straightforward to do.

Note that the formulations here clearly separate the governor controllers at the primary level and at the secondary level, while more conventional formulations combine the two levels into one PI controller that responds to average frequency changes ω; the proportional part comes from the primary level, and the integral part from the secondary. Physically these are two different control loops, and the formulation here for the first time provides two separate mathematical models for them, (5.42) and (5.125)–(5.129). A further comparison of the proposed formulation for secondary-level frequency control with ACE-based AGC implementation shows that the two formulations are actually consistent in terms of the measurement structure employed at this level. Note, however, that the need for defining the frequency bias for the ACE signal is entirely eliminated with the control scheme proposed here. The gain G_s is designed according to the desired frequency quality at different locations in the area ω_o. The gain H_s is a function of the transmission network parameters and is computed according to (5.137).

The same 5-bus example illustrated in Figure 5.3 is used here to illustrate the proposed control scheme. To simplify the computation, we choose all three generators to participate in secondary-level LFC control. In this case, $B_s = I$. The output variables are simply the three generator frequencies, so that $C_s = I$. The gain H_s is chosen according to (5.137). In this case, (5.138) is simplified to

$$\omega_G[k+1] = A_s\omega_G[k] - (I - \sigma D)G_s\omega_o^{set}[K] - \sigma D_P d_s[k] \qquad (5.140)$$

with $A_s = (I - \sigma K_P T_s) + (I - \sigma D)G_s$. To further simplify calculations, the gain G_s is chosen such that the three generator frequencies have decoupled identical dynamics, i.e., $A_s = \lambda I$ for some scalar constant λ. With this choice of A_s, one simply obtains

$$G_s = (I - \sigma D)^{-1}[(\lambda - 1)I + \sigma K_P T_s)] \qquad (5.141)$$

Figure 5.6 shows the same linearly varying load at bus #4 as in Figure 5.5 and the generator frequency responses to this load change with and without secondary-level LFC. The scalar constant λ is chosen to be $\lambda = -1$. Figure clearly indicates that the proposed secondary-level control scheme eliminates the situation of frequencies inversely following load variations, as depicted in Figure 5.5. The control scheme also significantly reduces the steady-state frequency errors.

5.5.2 Automated regulation of tie-line flows at the tertiary level

An important objective of tertiary-level control is to coordinate the regional secondary controllers in such a way that the interconnected system operates in an optimal fashion. To be more specific, the objective is to optimally reschedule the tie-line flows in response to the load variations $P_L[K]$ while maintaining the system frequency. At present, this task is carried out by agreement among several areas when needed. The concept is referred to in [6] for example, as "central AGC." The tertiary level is envisaged as particularly effective when the load generation mismatch in a specific subsystem exceeds the capacity of secondary-level controls, which only stabilize flows to their scheduled values. When certain control or output limits are reached in the stressed area, the scheduled exchange should be rescheduled to facilitate help from the neighboring areas. At present, this is done in an asynchronous, ad hoc manner. If this process were to become automated, it could be implemented in much the same way as digital AGC, only at a much slower sampling rate T_t. We observe that the presence of slow deviations in the average frequency documented in [6] can be explained by inadequate tie-line flow schedules.

The relevant output variables at the tertiary level are the generator frequencies and the tie-line flows. There exists a potential conflict between the setting of tie-line flows and the system frequency, because the tie-line flows and frequency are dependent on each other. An arbitrary setting for the tie-line flows could cause the system frequency to be offset from the desired value (60 Hz in the United States). This explains why in practice regional control is proportional to the time integral of a combination of the frequency deviation and the tie-line flow deviation, simply because when the tie-line flow setting is not properly chosen, neither of the deviations can be driven to zero. As a compromise, their combination is targeted to be vanished.

Conceptually, three types of implementation at the tertiary level can be envisioned: fully centralized, fully decentralized, and a combination of the two. The fully centralized approach entitles the tertiary-level decision maker to assign all set values for the feedback controllers so that a systemwide performance criterion is optimized. When the performance criterion is chosen as the total fuel cost needed to produce generation for meeting the anticipated load, the tie-line scheduling leads to what is known as security constrained economic dispatch. An advantage of our proposed control scheme is that it does not require a full information structure but only monitoring of critical load frequencies and tie-line flows, and it is therefore amenable to full automation.

The fully decentralized approach preserves the right of subsystems to schedule generation at their level in order to meet their own performance criteria, and it allows individual regions within the interconnected system to compete and make their own decisions. The dynamic properties of the rest of the system are not assumed to be known for the individual regions. In this case, each region measures the tie-line flows into it from the neighboring regions and, based on the measurements, determines the control strategies to optimize its own performance criterion. In this case, each region is completely independent and assumes no dynamic properties of the other regions.

The combined centralized/decentralized approach to tertiary-level scheduling preserves the decentralized structure of the control implementation but assumes structural properties of the other regions on the interconnected system. In other words, each region has knowledge of the dynamic structures of the other regions. Based on the shared knowledge of the dynamic structures of other regions and on-line measurements of the tie-line flows, each individual region make its decisions to optimize its own performance criterion. Intuitively, this approach would result in a systemwide performance between the fully centralized and fully decentralized implementations. This approach can be formulated into the framework of theoretical games.

In tertiary-control automation, one of the most important relations is the one relating the frequency schedules to tie-line flow schedules. The relation between the system power and the tie-line flows allows the proper setting of tie-line flows such that the system frequency is kept at the desired value and the system as a whole operates in an optimal fashion.

5.6 Summary

In this chapter, a structure-based modeling approach for real power/frequency dynamics of an interconnected power system is presented. Dynamics of the system are formulated by combining the local dynamics of individual generators and the network couplings. The concept of structural singularity for large power systems is defined. It is shown that the decoupled real power/frequency dynamics of power systems are structurally singular. Discrete-time dynamic models on slower time scales are derived. The modeling approach provides a solid basis for systematic control design.

This chapter also presents a new hierarchical control design concept for the real power/frequency dynamics of power systems, based on the structural models developed in Chapter 4. Systemwide frequency regulation on longer time scales is introduced. Simulations show that the new control scheme significantly reduces frequency fluctuations in the presence of load variations. A tabular representation of performance objectives at the three level of hierarchy (Figure 3.3) is shown in Table 5.3.

Table 5.3: Generalized AGC/LFC scheme

Control level	Goal (objective)	Control function	Constraints	Relevant information
Tertiary	Given P_{ij}^{max}, provide AGC/LFC supplementary signals $\omega_{Gi}^{ref}[KT_t]$ according to $min \sum_{i=1}^{N_A} c_i P_{Gi}$	Provide $\omega_{Gi}^{ref}[KT_t]$ $i = 1, \cdots, N_A$ $F_{Gi}[KT_t]$ as T_{ij} occurs	Tertiary level aggregate model relating $F_G[KT_t]$, $F_L[KT_t]$, $\omega_G^{ref}[KT_t]$ $P_G[KT_t]$ $F_{Gi} + P_{Gi} = P_G^N$ $F_{Li} - P_{Li} = P_L^N$	input: P_{ij}^{max} output: $\omega_{Gi}^{ref}[KT_t]$ for all $i \in N_A$ feedback signals: all $ACE_i[KT_t]$
Secondary	Given P_{ij}^{max}, provide AGC/LFC supplementary signal $\omega_{Gi}^{ref}[KT_t]$ while keeping exchange with other regional fixed	Provide $\omega_{Gi}^{ref}[KT_t]$ to accommodate P_{ij}^{max} and assure $\Delta F_{Gi} = 0$	$\omega_{Gi}^{min} < \omega_{Gi} < \omega_{Gi}^{max}$	input: P_{ij}^{max} output: $\omega_{Gi}^{ref}[KT_s]$ feedback signals: $ACE_i[KT_t]$ $i = 1, \cdots, k$
Primary governors (on all generat- -ing units)	Provide stable frequency response within $t < T_s$ for the given range $P_{Gi}^{min} < P_{Gi} < P_{Gi}^{max}$	Stabilize P_{Gi} according to its droop characteristics for given $\omega_{Gi}^{ref}[kT_s]$	$P_{Gi}^{min} < P_{Gi} < P_{Gi}^{max}$ $\omega_{Gi}^{min} < \omega_{Gi} < \omega_{Gi}^{max}$ $\omega_{Gi}^{ref\,min} < \omega_{Gi}^{ref} < \omega_{Gi}^{ref\,max}$	input: $\omega_{Gi}^{ref}[KT_s]$ outputs: ω_{Gi}, P_{Gi} feedback signal: $(\omega_{Gi}^{ref}[kT_s] - \omega_{Gi})$

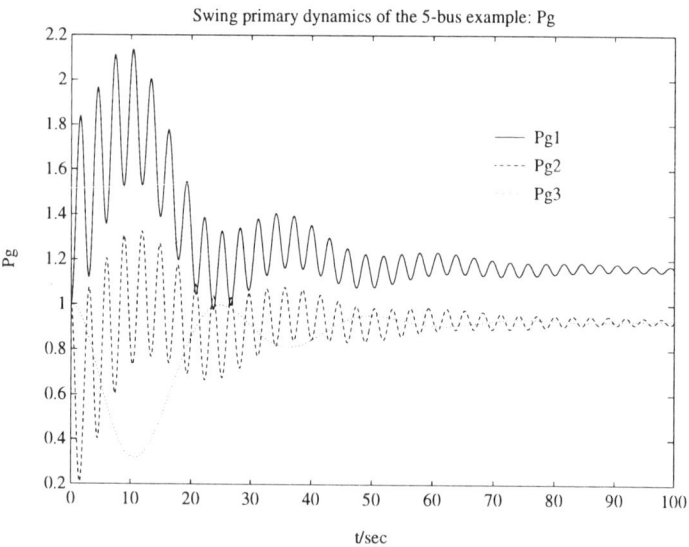

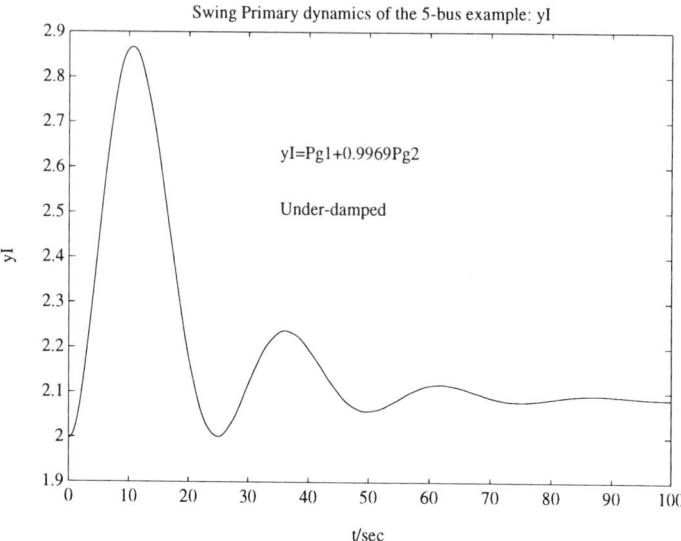

Figure 5.4: Inter-area oscillations

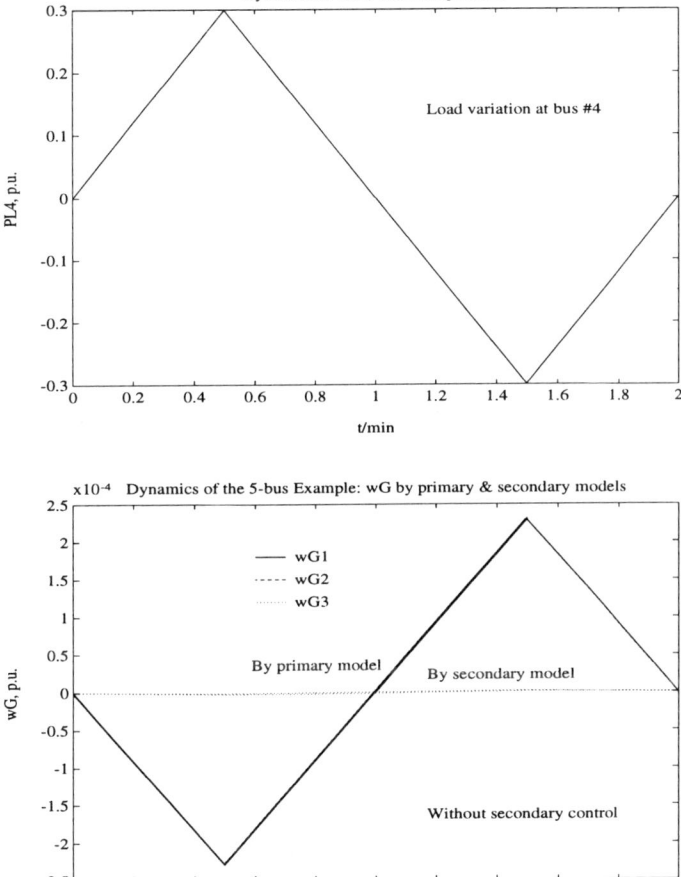

Figure 5.5: Load variation and frequency response

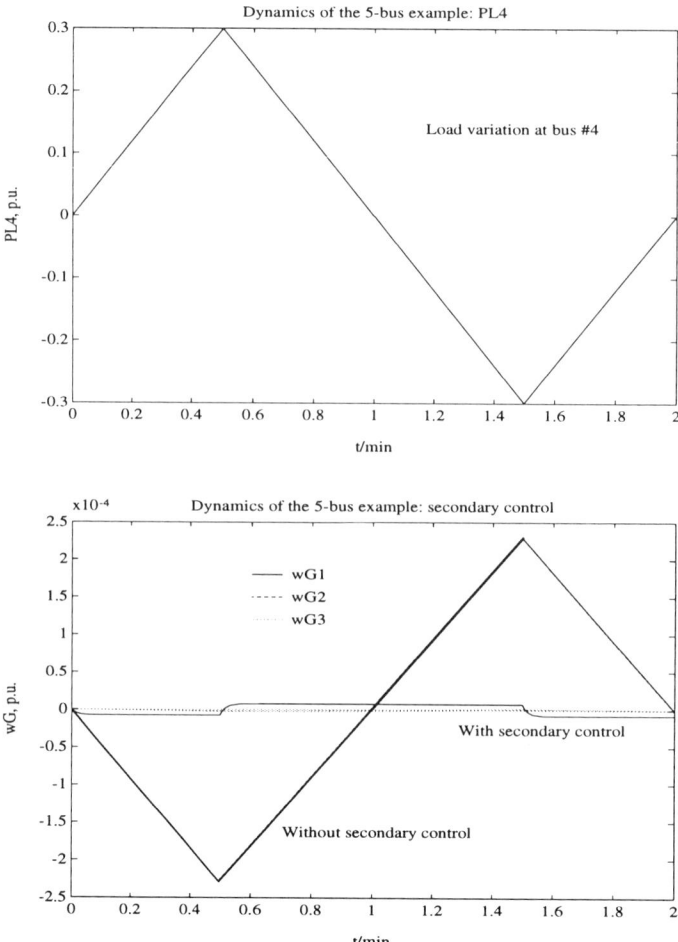

Figure 5.6: Load variation and frequency response

CHAPTER 6
GENERATION-BASED REGULATION OF REACTIVE POWER/VOLTAGE

The work on reactive power/voltage dynamics presented in this chapter is motivated by the practical need to regulate load voltages over mid- and long-term horizons according to a specified performance criterion by changing generator voltage settings. While the voltage control of an interconnected large-scale power system is widely recognized as a very important problem, its basic formulation and solutions are often utility-specific. Most often the voltage control is viewed as an entirely static problem, whose solution is identical to centralized open-loop optimization-based reactive power/voltage management. The most common tool for solving this optimization problem is an algorithm based on optimal power flow (OPF). This approach computes changes in generator voltages needed to regulate load voltages on the entire interconnected system. It assumes the availability of full systemwide information and requires a large amount of data, typically not available when the system is experiencing unusual operating conditions. Moreover, this approach does not provide an easy balance between coordination and competition, which is of practical interest in the coordination of regions.

A second approach to voltage control coordination relies on decomposition of a large system into regions and an on-line, decentralized, closed-loop, reduced information structure for controlling regions. For instance, the French power system has been committed to full automation of systemwide voltage regulation while employing an intuitive reduced information structure at the regional level. The dynamic model of voltage over mid- and long-term horizons, as developed in this chapter, is a load variation and control-driven model, in the sense that any variations in voltages with time are caused only by the control signal or the disturbances. It will be shown that for this kind of control-driven model, only a certain number of states can be fully controlled. The maximum number of states that can be fully controlled is equal to the number of controls. Therefore only at most the same number of load voltages as generators can be independently controlled. These states chosen to be independently controlled are referred to as the *pilot load voltages*. The pilot load voltages are controlled within each region by regional controllers, and it is assumed that neighboring regions have negligible effects. In this case, the responsibility for coordinated voltage regulation is shared among regional closed-loop controllers (the secondary voltage controllers) and the operators at the national control center (the tertiary level).

The work reported in this chapter was largely sponsored by Electricité De France (EDF). As the French power network has become increasingly meshed during the past decade and is operated closer to the prespecified voltage limits, tertiary-level coordination of the regional secondary voltage controllers has become critical to improve the security and economics of the entire system. Under mild system changes, it is often sufficient to rely on the operator's expertise to provide coordinating signals from the national level. If time permits and if the results of the state estimation are available, the operator could employ computer tools such as OPF to assist in the decision making. However, when the interconnected network experiences unusual reactive power deficiency, typically in one region at a time, it is important to provide the dispatchers quickly with adequate coordination strategies.

The main goal here is to develop new concepts for coordination of secondary voltage controllers at the national level that preserves a reduced, pilot point–based information structure. As in the frequency dynamics case, a structurally-based decomposition approach is taken to derive the regional voltage dynamic model consisting of local dynamics of individual devices and network coupling. A remarkable feature is that although the reactive power/voltage control has a parallel structure to the real power/frequency control, these two models represent fundamentally different systems. As shown in Chapter 5, the real power/frequency dynamics are structurally singular. This is, however, not the case for reactive power/voltage dynamics, unless a specific numerical singularity occurs. The structural singularity of real power/frequency dynamics results from the fact that real power across a transmission line depends on the phase angle difference across the line only. A constant offset on both angles makes no difference for real power transfer. However, reactive power across a transmission line is not a function of the voltage difference only, so that a constant offset on both voltages *does* change the reactive power flow.

On the other hand, a different type of structural singularity occurs at the secondary level with quasi-static control-driven voltage models because of an insufficient number of controls. Interaction variables associated with this type of structural singularity are defined as *quasi-static interaction vari-*

ables. Both the continuous interaction variables of frequency dynamics and the quasi-static interaction variables of voltage dynamics represent singular modes of the system evolving at different time scales. The most important property of these two kinds of singularities is that they are structurally inherent rather than numerically coincident, as already shown for the case of frequency dynamics. These structural singularities on different time scales reveal fundamental dynamic properties of power systems. An important breakthrough as a result of structural decomposition is that the interaction variables defined here represent physically measurable variables, such as reactive tie-line flows. This provides a basis for automated feedback control design because interaction variables are physically measurable.

It should be emphasized that, in contrast to many standard techniques on system decomposition that assume weak or sparse coupling among the regions, the structural approach developed in this chapter does not depend on this assumption, as already seen in the case of real power/frequency dynamics. The decomposition is entirely based on the structural properties of the system. In the particular case of weak interconnections, the structural interaction variables defined here can be shown to represent the slow dynamics obtained using the singular perturbation method.

6.1 Modeling

Having introduced structurally based models for real power/frequency control at different levels of hierarchy in the previous chapters, we review the present state of the art in voltage control via generators. It is shown here that an effective formulation of voltage control at all levels lends itself naturally to exploiting the same structural properties as in the design of real power/frequency controls of large electric power systems. This fact appears not to have been recognized in the literature, mainly because of the somewhat inaccurate notion that a single average frequency can be associated with each subsystem, while voltages could be quite non-uniform throughout a voltage control area. The need for more refined frequency measurements and specifications throughout each subsystem is described in the previous chapters, and a model formulation that unbundles frequencies at the locations of interest is proposed as more realistic, particularly when considering regulatory changes imposed on operating power systems in the future. Depending on electric characteristics of each subsystem non-uniformity in frequency deviations could be significant. Eq. (5.32) explicitly defines how frequencies vary at specific locations in the system as a function of real power generation and demands.

The second fundamental issue in comparing the concepts for frequency and voltage control automation is the issue of reactive power losses, which are much larger than real power losses. A look at the systems-oriented literature in the area of real power/frequency dynamics reveals that the most elegant derivations are proposed under the assumption of no real power losses. It is expected that attempts to extend such existing formulations for real power dynamics to reactive power will not succeed. It is with this issue in mind that we emphasize that the formulation proposed here does not make any assumptions regarding losses, i.e., the linearized real and reactive power flows are not assumed to be lossless.

The concept proposed in this chapter is not dependent on the above two "facts," which create potential qualitative differences in control concepts for real and reactive power. We suggest a structurally based concept for voltage control automation that is parallel to the real power/frequency control introduced in Chapter 5. In that chapter, we established a structural approach to support automated load frequency control (LFC), already operating. Yet, while primary-level voltage controllers are fully automated, secondary-level voltage automation based on a principle parallel to LFC is being implemented in France and Italy only. Moreover, tertiary-level voltage automation has not yet been implemented anywhere in the world. We propose, first, that secondary-level voltage control automation can be improved relative to the existing implementations in France and Italy [103, 63], by taking into account the effects of neighboring areas at each subsystem level. Second, tertiary-level voltage coordination can be fully developed following a general parallel with automatic generation control (AGC).

To establish a background for automatic voltage control, let us recognize that the structural settings for deriving models of voltage dynamics are identical to the ones exploited in real power dynamics, illustrated in Fig. 5.1. Primary voltage dynamics defined by the electromagnetic changes at each generator and the excitation system controls can be expressed in the same structural form as the frequency dynamics in (5.42) by introducing local state variables. These local state variables define the local primary voltage dynamics of each generator/excitation system in response to terminal voltage deviations from its reference value.

6.1.1 Local dynamics

The general structure of reactive power/voltage control of power systems is shown in Figure 6.1, and a typical excitation system is shown in Figure 6.2.

The excitation system controls the field voltage input to the generator using the error signal between the measured generator terminal voltage and a given reference value. The goal is to maintain the generator terminal voltage at the prespecified setting.

As in the real power/frequency control case, the dynamics of all the generators are coupled through the transmission network via the reactive power outputs of all generators. The transmission network imposes an algebraic constraint on the total reactive power injections into the network, and couples all the generator power outputs. It also couples generators to the loads.

Chapter 6	123

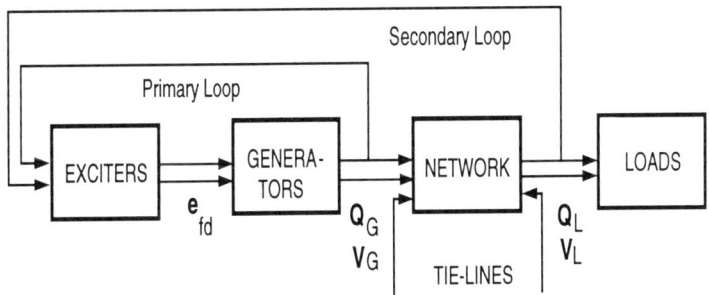

Figure 6.1: Typical structure of reactive power/voltage control

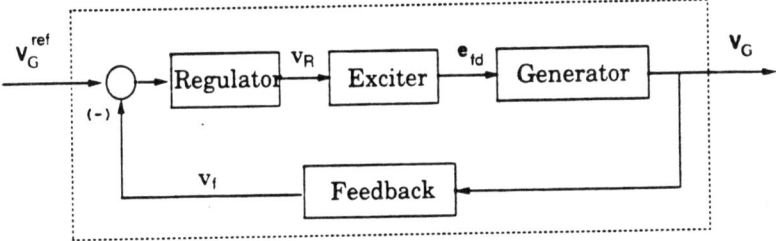

Figure 6.2: A typical excitation system

Let us first consider the local dynamics of an individual excitation system and generator set. A schematic block diagram for the excitation control system is shown in Figure 6.2.

A typical excitation system consists of the regulator, the exciter, and the excitation feedback compensator. The regulator is described by

$$T_a \dot{v}_R = K_a v_f - \frac{K_a K_f}{T_f} e_{fd} - v_R - K_a(V_G - V_G^{ref}) \tag{6.1}$$

where e_{fd} is the field voltage of the exciter, V_G is the generator terminal voltage, and V_G^{ref} is the reference value for the generator terminal voltage.

The exciter dynamics can be modeled in the following form

$$T_e \dot{e}_{fd} = -(K_e + S_e)e_{fd} + v_R \tag{6.2}$$

Generator dynamics are typically given by

$$T'_{do}\dot{e}'_q = -e'_q - (x_d - x'_d)i_d + e_{fd} \tag{6.3}$$

neglecting the effects of damper winding, i.e., $e'_d = 0$, where i_d is the reactive current out of a generator. Current component i_d is also a function of the states, known as the network constraints. Under the assumption $e'_d = 0$, we obtain

$$i_d = \frac{Q_G}{e'_q} \tag{6.4}$$

since $Q_G = e'_q i_d - e'_d i_q$.

Finally, the compensator is of the form

$$T_f \dot{v}_f = -v_f + \frac{K_f}{T_f} e_{fd} \tag{6.5}$$

Define the local states of each generator as

$$x_{LC} = [v_R \; e_{fd} \; e'_q \; v_f]^T \tag{6.6}$$

One can write (6.1)–(6.5) together in a nonlinear state space form as

$$\dot{x}_{LC} = f_{LC}(x_{LC}, Q_G, V_G^{ref}) \tag{6.7}$$

recognizing that $V_G = e'_q$, since $V_G = \sqrt{e'^2_q + e'^2_d}$ and $e'_d = 0$. This model of local dynamics is of the general form (4.9). In this model, the generator reactive power output Q_G is the coupling variable x_{CP} in the general local dynamics model (4.9). This variable couples dynamics of all generators connected through the network. This coupling, known as the network constraints, is discussed next.

6.1.2 Network constraints

As in the real power/frequency case, generator reactive power outputs Q_G are determined by the interactions with other generators and loads via the transmission network. Derivations here are very similar to those for the real power/frequency dynamics. Consider any administrative region with m generators and n loads within an interconnected system. The complex power constraint for the transmission network is stated in (5.21). The real part of the equation, i.e., the real power/frequency constraint, was discussed in Chapter 5. The imaginary part is the reactive power/voltage constraint. Similar to (5.22), let us write the imaginary part of (5.21) as

$$Q^N = Q^N(\delta, V) \tag{6.8}$$

where

$$\delta \triangleq \begin{bmatrix} \delta_G \\ \delta_L \end{bmatrix} \quad \text{and} \quad V \triangleq \begin{bmatrix} V_G \\ V_L \end{bmatrix} \tag{6.9}$$

are the nodal voltage angles and magnitudes. Define the reactive power tie-line flows from the neighboring areas into all generator nodes and all load nodes as

$$F_G \triangleq \begin{bmatrix} F_G^1 \\ \vdots \\ F_G^m \end{bmatrix} \quad \text{and} \quad F_L \triangleq \begin{bmatrix} F_L^1 \\ \vdots \\ F_L^n \end{bmatrix} \tag{6.10}$$

Note that we use F_G and F_L to represent the reactive power tie-line flows. Then it is obvious that $Q_G^N = F_G + Q_G$. The network constraints for the reactive power balance can be further written as

$$Q_G^N(\delta, V) = F_G + Q_G \tag{6.11}$$

The same relation is true for the load buses, i.e.,

$$Q_L^N = F_L - Q_L \tag{6.12}$$

and the network constraints (6.8) at the load nodes are expressed as

$$Q_L^N(\delta, V) = F_L - Q_L \tag{6.13}$$

assuming again that the positive direction for tie-line flows is the injection into the network, and the positive direction for loads is leaving the network. As with the real power/frequency case, we separate reactive power injections into the part of the injection from the generators or loads and the part from the interconnecting tie lines with neighboring regions.

These represent algebraic constraints to the network variables like bus voltage vector V. They add severe difficulties to the local dynamic models in differential equations. To eliminate the algebraic constraints, we differentiate on the algebraic constraints. Here the real/reactive power decoupling assumption is made again. In terms of reactive power relations, the decoupling assumption is expressed as

$$\partial Q^N / \partial \delta = 0 \tag{6.14}$$

Under this assumption, differentiating the above constraints yields

$$\dot{F}_G + \dot{Q}_G = J_{GG} \dot{V}_G + J_{GL} \dot{V}_L \tag{6.15}$$

$$\dot{F}_L - \dot{Q}_L = J_{LG} \dot{V}_G + J_{LL} \dot{V}_L \tag{6.16}$$

where

$$J_{ij} = \frac{\partial Q_i^N}{\partial V_j}, \quad i,j = G, L \tag{6.17}$$

are the Jacobian matrices evaluated at the given operating point. Assuming J_{LL} to be invertible under the normal operating conditions, we define one of the most important matrices associated with a transmission network

$$C_V = -J_{LL}^{-1} J_{LG} \tag{6.18}$$

to express voltage deviations at loads V_L in terms of voltage deviations at generators V_G and fluctuations in load power. It follows from (6.16) that

$$\dot{V}_L = C_V \dot{V}_G + J_{LL}^{-1}(\dot{F}_L - \dot{Q}_L) \tag{6.19}$$

where

$$V_G \triangleq \begin{bmatrix} V_G^1 \\ \vdots \\ V_G^m \end{bmatrix} \tag{6.20}$$

Relation (6.19) defines the explicit dependence of load voltages on generator voltages determined by the network constraints. Combining (6.19) and (6.15), and defining the other two most important matrices associated with a transmission network

$$K_Q \triangleq J_{GG} + J_{GL} C_V \tag{6.21}$$

and

$$D_Q \triangleq -J_{GL} J_{LL}^{-1} \tag{6.22}$$

result in

$$\dot{Q}_G = K_Q \dot{V}_G - \dot{F}_e + D_Q \dot{Q}_L \tag{6.23}$$

Here F_e represents effective reactive tie-line flow as seen by each generator and is defined as

$$F_e \triangleq F_G + D_Q F_L \tag{6.24}$$

Eq. (6.23) defines the relation among the generator reactive power outputs Q_G, the generator voltages, the tie-line flows into the subsystem, and the reactive load variations, through the network characteristics specified by the two important matrices K_Q and D_Q.

It should be noted that any (portion of a) network is fully characterized by the three matrices (K_Q, C_V, D_Q), with K_Q reflecting the effect of the generator frequencies on the generator real power outputs, C_V relating the generator frequencies to the load frequencies, and D_Q representing different electrical distances of loads at different locations seen by the generators. It is easily shown that matrix K_Q does not have the structural singularity associated with matrix K_P for the real power/frequency dynamics, although under extreme operating conditions numerical singularities are possible.

6.1.3 Structural dynamic model

Local dynamic model (6.7) and the network relation (6.23) combine to form the structural dynamic model for reactive power/voltage of the administrative region. Define the state variables of the region as in (4.18). Voltage dynamics of the region can be written in a standard nonlinear state space form as

$$\dot{x} = f(x, V_G^{ref}, \dot{F}_e) \qquad (6.25)$$

where the nonlinear function f is a combination of (6.7) and (6.23). This model is of the general form (4.20).

In contrast to the real power/frequency dynamics, the network matrix K_Q is in general not singular. As a result, the voltage dynamics as described in (6.25) are not structurally singular. Because of this nonsingularity, inter-area voltage oscillations in general do not exist, except in the case of numerical singularities. Also there is in general no need for direct tie-line flow control, as in the frequency case to remove the structural singularity, and the interaction variables as defined there do not exist. However, as will be discussed next, there exists another kind of quasi-static interaction variables on the secondary level, which reflect a different type of structural singularity—insufficient number of controls for the quasi-static voltage dynamic model.

6.2 Quasi-static voltage model

Typical designs of the excitation system yield very fast transient dynamics, relative to the secondary time scale T_s. A quasi-static voltage dynamic model can be derived when the reference value V_G^{ref} is updated at discrete time instant kT_s. First set all derivatives in (6.25) equal to zero. This leads to an algebraic equation

$$f(x, V_G^{ref}, 0) = 0 \text{ at } kT_s \qquad (6.26)$$

The linearized equation for this is

$$A\delta x + B\delta V_G^{ref} = 0 \text{ at } kT_s \qquad (6.27)$$

where A is nonsingular. This nonsingularity determines a unique relation between the generator voltage V_G, which is part of the state variables, and the reference value V_G^{ref}. For simplicity, let us simply write

$$\delta V_G[k] = \alpha \; \delta V_G^{ref}[k] \tag{6.28}$$

In other words, generator terminal voltages are directly proportional to the reference values.

The secondary-level control of reactive power/voltage is to regulate load voltage profiles with generator terminal voltages. To derive the relation between load voltages and generator terminal voltages, let us consider any administrative region within an interconnected system. From the network relation (6.19) one can obtain the following quasi-static discrete-time model, by integrating (6.19) from the secondary time instant kT_s to the next instant $(k+1)T_s$,

$$V_L[k+1] - V_L[k] = C_V(V_G[k+1] - V_G[k]) + J_{LL}^{-1}[(F_L[k+1] - F_L[k])$$

$$-(Q_L[k+1] - Q_L[k])] \tag{6.29}$$

As in the frequency control case, we define the secondary corrective control signal as

$$u_s[k] \triangleq V_G[k+1] - V_G[k] \tag{6.30}$$

the tie-line flow changes as

$$f[k] \triangleq F_L[k+1] - F_L[k] \tag{6.31}$$

and the secondary-level load disturbances as

$$d_s[k] \triangleq Q_L[k+1] - Q_L[k] \tag{6.32}$$

Eq. (6.29) then becomes

$$V_L[k+1] - V_L[k] = C_V u_s[k] + D_s(f[k] - d_s[k]) \tag{6.33}$$

with $D_s \triangleq J_{LL}^{-1}$. This is the desired secondary-level discrete-time dynamic model for a region within an interconnected system.

To write this in standard control notation, let us define the secondary-level states as

$$x_s[k] \triangleq V_L[k] \tag{6.34}$$

Then (6.33) is rewritten as

$$x_s[k+1] - x_s[k] = C_V u_s[k] + D_s(f[k] - d_s[k]) \tag{6.35}$$

where $f[k]$ and $d_s[k]$ act as the disturbances to the system with control $u_s[k]$. The difference between $f[k]$ and $d_s[k]$ is that the flow can be measured, while the loading variations are rarely measured in practice. They are either estimated or simply treated as a real disturbance.

It is emphasized that the discrete-time dynamic model (6.35) is a control- or disturbance-driven model, in the sense that, if all disturbances (including tie-line flows) are not present, and the corrective control is inactive, then $x_s[k] = $ constant, $\forall k$. Clearly, control actions are needed to bring the system (6.35) back to the nominal operation if the system is perturbed away from the nominal operation by disturbances.

It is also worth noting that the secondary-level control is an implicit integral control. We see from (6.30) that the secondary-level control $u_s[k]$ is a corrective signal, i.e., it updates the previous generator voltage $V_G[k]$ to get the next value. This corrective action is equivalent to an integral control. It is this implicit integral control that can eliminate steady-state errors in the load voltages caused by system disturbances.

6.3 Quasi-static interaction variables

The secondary quasi-static voltage dynamic model for any region explicitly in terms of the tie-line flows has been derived as

$$x_s[k+1] - x_s[k] = C_V u_s[k] + D_s(f[k] - d_s[k]) \tag{6.36}$$

where $x_s[k] = V_L[k]$ is the state vector consisting of all load voltages. The dimension of the sensitivity matrix C_V is $n \times m$, where n is the number of load buses, and m is the number of generator buses that participate in the secondary-level regulation. In general, it is true that $n > m$, i.e. the number of load buses is larger than the number of generator buses participating in secondary-level control.

Under the condition of $n > m$, one can verify that the closed-loop system using any feedback control is singular, because matrix C_V has maximum rank of m. This structural singularity is due to the relative numbers of controls and states. This is a general property for any control-driven system. In exploiting this structural singularity of the quasi-static voltage dynamics, we first give the following definition.

- **Definition 6.1 (Quasi-Static Interaction Variables):** Any linear combination of the states, $z[k] = T x_s[k]$, $T \neq 0$, that satisfies

$$z[k+1] - z[k] \equiv 0, \quad \forall k \tag{6.37}$$

for any secondary control actions, and in the absence of interactions among regions and the disturbance, i.e., $f = 0$ and $d_s = 0$, is defined as the quasi-static interaction variable of the administrative region under study.

The same notation as with the continuous interaction variables is used here to indicate the similar characteristics of the two. The meaning is clear from the context under study. As with the continuous interaction variables, the quasi-static interaction variables do not vary with time when interconnections are removed and load disturbances are not present. For the interconnected system, therefore, any variations of the interaction variables with time are entirely due to the interactions among regions or load disturbances. Note from the definition that the interaction variables are not unique. In fact, any combinations of the interaction variables are still interaction variables.

Let us derive the condition for the transformation matrix T. Combining (6.37) and (6.36) yields

$$z[k+1] - z[k] = T(V_L[k+1] - V_L[k]) = TC_V u_s[k] + TD_s(f[k] - d_s[k]) \quad (6.38)$$

Under the conditions in the definition, $f[k] \equiv 0$ and $d_s[k] \equiv 0$, we arrive at

$$z[k+1] - z[k] = TC_V u_s[k] \quad (6.39)$$

In order to have $z[k+1] - z[k] \equiv 0$ for any control $u_s[k]$, matrix T must satisfy

$$TC_V = 0 \quad (6.40)$$

This is the desired equation for calculating T. Note that matrix C_V has maximum rank $m < n$, and therefore (6.40) has nonzero solutions for T. We can solve T from (6.40), since it is a simple algebraic equation, which can be solved with a Gaussian elimination method. The need for eigenstructure analysis is completely avoided.

Note that the definition for interaction variables is for any secondary control, meaning that the interaction variables are independent of the specific secondary control. Equivalently, the secondary control cannot affect the interaction variables. Any variations of the interaction variables are due to the interactions with other regions or the load variations. The matrix T, as a result, will not be dependent on the specific form of the secondary control.

An interesting difference between the continuous and quasi-static interaction variables can be noted. For any single region, the dimension of the continuous interaction variables is in general 0, because the system normally has a full rank. For the quasi-static interaction variables, as shown above, the dimension is the difference between the number of states and the number of controls. This difference between the interaction variables indicates the fundamentally different causes for the two types of interaction variables.

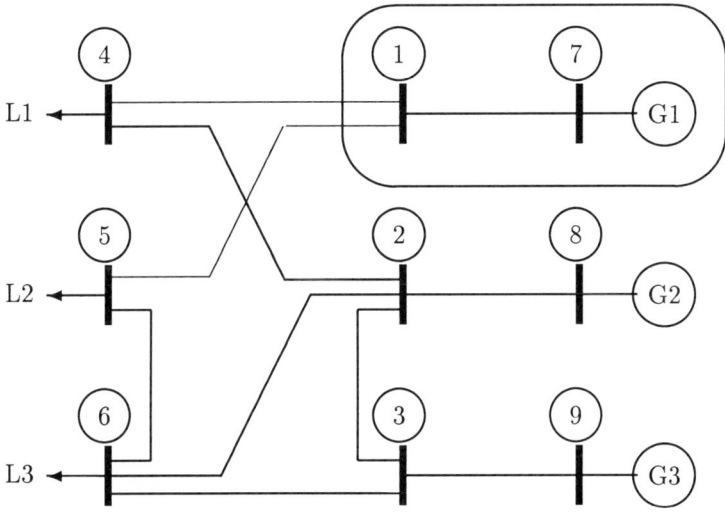

Figure 6.3: The 9-bus example

Once the interaction variables are determined from (6.40), one can further derive the dynamic model for these interaction variables. Eqs. (6.38) and (6.40) lead to

$$z[k+1] - z[k] = TD_s(f[k] - d_s[k]) \qquad (6.41)$$

This is the desired dynamic model for the interaction variables. This simple model relates the interaction variables to the tie-line flows and load variations. It is of crucial importance for secondary-level voltage control and tertiary-level coordination, as discussed in more detail in Chapter 7.

Notice that the definition for the quasi-static interaction variables does not assume numerically weak interconnections. Rather, it reflects a structural property of the system, different numbers of states and controls. It is interesting to relate the interaction variables defined above to the slow variables in singular perturbation analysis when the interconnections are indeed weak. It can be seen from the interaction dynamic model (6.41) that, in the weak interconnection case, the interaction variables do vary more slowly than the rest of the states. One can rigorously prove that, in the weak interconnection case, the interaction dynamics derived here will be the slow subsystem in the singular perturbation analysis.

Let us now demonstrate the interaction variables and their properties by a small 9-bus example network, given in Figure 6.3.

Region I consists of buses #1 and #7; the rest is region II. The pilot points are buses #1, #2, and #3. The feedback gain G_s is designed such that the pilot voltages settle exponentially in 3 minutes. The load is assumed to have a step increase at bus #5 at $t = 0$; thus the effect of $d_s[k]$ is seen in changes of the initial conditions for all the load voltages. This example will be used throughout to demonstrate the developed concepts. Simulations for the EDF network will also be given, but only for the purpose of illustrating tertiary-level control. The numerical data used in the simulations for the 9-bus system are given in Table 6.1.

					Line Parameters					
		#1-4	#1-5	#1-7	#2-3	#2-4	#2-6	#2-8	#3-6	#3-9	#5-6
b		5	7	7.69	15.67	21.55	7.19	8.33	12.49	10	9.8
g		0	.15	0	0	1.02	.34	0	.31	0	0
					Nominal Operating Point					
		#1	#2	#3	#4	#5	#6	#7	#8	#9	
V		1	1	1	1	1	1	1	1	1	
δ		0	0	0	0	0	0	0	0	0	

Table 6.1: Per-unit data of the 9-bus example

The sensitivity matrix for region I, which has only one generator bus and one load bus, is simply

$$C_V^I = 1 \tag{6.42}$$

The sensitivity matrix for region II, which has two generator buses and five load buses, is calculated as

$$C_V^{II} = \begin{bmatrix} .55 & .45 \\ .37 & .63 \\ .55 & .45 \\ .44 & .56 \\ .44 & .56 \end{bmatrix} \tag{6.43}$$

Let us use condition (6.40) to calculate the matrix T. For region I, condition $T^I C_V^I = 0$ simply gives $T^I = 0$. This means that region I does not have any interaction variables. For region II, condition $T^{II} C_V^{II} = 0$ leads to one independent solution

$$T^{II} = \begin{bmatrix} 21.55 & 0 & -21.55 & 0 & 0 \\ 0 & 0 & 0 & 9.8 & -9.8 \\ 7.19 & 12.49 & 0 & 9.8 & -29.48 \end{bmatrix} \tag{6.44}$$

There are three independent interaction variables given by

$$z^{II} = \begin{bmatrix} 21.55(x_2 - x_4) \\ 9.8(x_5 - x_6) \\ 7.19 x_2 + 12.49 x_3 + 9.8 x_5 - 29.48 x_6 \end{bmatrix} \tag{6.45}$$

To see the physical meaning of these interaction variables, let us rewrite the above equation as

$$z^{II} = \begin{bmatrix} 21.55(x_2 - x_4) \\ 9.8(x_5 - x_6) \\ 7.19(x_2 - x_6) + 12.49(x_3 - x_6) + 9.8(x_5 - x_6) \end{bmatrix} \quad (6.46)$$

Note that the interaction variables are given in the differences of bus voltages, with the coefficients being exactly the line inductances. Therefore, these interaction variables represent the power flows on the lines, since the power flow on each line is exactly the voltage difference across the line multiplied by the line inductance, for the nominal operating conditions given in Table 6.1. Preservation of physical meaning of the interaction variables is important for both regional control and tertiary coordination.

6.4 Voltage regulation

As discussed above, voltage regulation of the power system involves the excitation system to stabilize the generator terminal voltages to their given reference values. These reference values are adjusted at discrete time instants slowly by the higher-level controls. Much effort has been given to analysis, modeling, and design of excitation systems. This chapter will not further discuss issues concerning the design of fast primary controls. Emphasis here is on higher-level control design—the slow updating of the reference values for excitation systems. The ultimate goal is to develop reliable automated regional and systemwide voltage control to enhance secure and economical operation of power systems.

6.5 Regional voltage control

The simple control-driven model (6.35) is basic to developing decentralized secondary-level voltage controllers. This control level is referred to in France and Italy as automatic voltage control (AVC) [103, 101]. Its main function is to respond to reactive load disturbances $d_s[k] = Q_L[k+1] - Q_L[k]$. AVC is implemented on generator units whose voltage set points $V_G^{ref}[k]$ are automatically changed to respond to deviations in load voltages $V_L[k]$ at the chosen subset of loads, the critical pilot point loads, $V_c[k] = C_s V_L[k]$. Because the maximum number of states that can be fully controlled is equal to the number of active controls (the number of participating generators), the number of pilot point loads is always less or equal to the number of participating generators.

As with AGC, AVC at the secondary level should be designed to keep operation of subsystems as autonomous as possible, given control constraints.

Its main objectives are to

- Reschedule $V_G^{ref}[k]$ at each subsystem level to meet reactive load deviations $Q_L[k]$, $k = 0, 1, \ldots$.
- Regulate voltages $V_c[k]$ to $V_c[K]$.
- Maintain $F[K] \approx 0$ as long as reserves within each area are available (area control principle).
- Optimize subsystem performance (total reactive reserve or total transmission losses).

The formulation offered in this chapter enables one to perform optimization of a chosen performance criterion in a coordinated manner with voltage regulation at the pilot point loads. It follows that the secondary-level control law of general form (4.39) or (4.42) will achieve the above objectives. It is important to recognize that the "participation factors" of different units are directly determined by the optimal gain G_s for any given power network; no additional economic dispatch functions are required at a subsystem level.

Consider, within an interconnected system, an area with n load buses and m generators participating in secondary-level control. The dynamic equation for all the load voltages is given by

$$V_L[k+1] - V_L[k] = C_V u_s[k] + D_s(f[k] - d_s[k]) \tag{6.47}$$

where the dimension of the sensitivity matrix C_V is $n \times m$. In general, $n > m$. Under this condition, one can prove the following:

- **Proposition 6.1 (Controllability):** The dynamic system given in (6.36) is not fully controllable.

Proof: Let us write the controllability matrix for (6.36) as

$$Q_c = \begin{bmatrix} B & AB & \cdots & A^{n-1}B \end{bmatrix} = \begin{bmatrix} C_V & 0 & \cdots & 0 \end{bmatrix} \tag{6.48}$$

since $A = 0$ for the control-driven system. Because the sensitivity matrix C_V has maximum rank m, the system is not fully controllable.

As a result, not all load voltages can be fully controlled. Only the same number of loads as the number of generators can be fully controlled. This leads to the idea of pilot load voltages, the number of which does not exceed the number of generators participating in secondary-level control.

Because of this controllability issue, the maximum number of loads that can be controlled is equal to the number of active controls. Let us choose m output variables, or pilot loads, as

$$V_c[k] = C_s V_L[k] \tag{6.49}$$

The control-driven dynamic model for these pilot loads can be obtained as

$$V_c[k+1] - V_c[k] = C_c u_s[k] + D_c(f[k] - d_s[k]) \qquad (6.50)$$

where $C_c = C_s C_V$ is the sensitivity matrix of the critical pilot voltages relative to the control, and $D_c = C_s D_s$. This is the basic model for secondary-level control design. Note that C_c is a square matrix.

6.5.1 Conventional secondary-level control

The present state of secondary-level voltage regulation is based only on regional measurements, i.e., regional pilot point loads. The effect of interconnecting flow changes due to changes in the neighboring regions is not considered directly. One consequence of this is that under certain conditions the secondary controller may cause a significant overshoot or not reach the set value within the prespecified time intervals.

The goal of secondary-level regulation is to maintain the pilot load voltages (output variables) at their prespecified set values when the system is under disturbances. A simple proportional feedback law takes the form

$$u_s[k] = G_s(V_c[k] - V_c[K]) \qquad (6.51)$$

where $V_c[K] \equiv V_c(KT_t)$ is the value for the pilot load voltages on the tertiary-level time scale. This value is adjusted by tertiary-level control on the even longer time scale T_t, and is a constant for the secondary-level control process (see Figure 3.4).

Under this control law, the secondary-level closed-loop dynamic model becomes

$$V_L[k+1] - V_L[k] = C_V G_s(C_s V_L[k] - V_c[K]) + D_s(f[k] - d_s[k]) \qquad (6.52)$$

and the pilot point voltage dynamics are

$$V_c[k+1] - V_c[k] = C_c G_s(V_c[k] - V_c[K]) + D_c(f[k] - d_s[k]) \qquad (6.53)$$

Let $A_s = C_c G_s$. We then rewrite (6.52) as

$$V_c[k+1] - V_c[k] = A_s(V_c[k] - V_c[K]) + D_c(f[k] - d_s[k]) \qquad (6.54)$$

One requirement for the choice of pilot load voltages V_c is that the resulting matrix C_c must be nonsingular. If matrix C_c is singular, then A_s will be singular for any gain matrix G_s. The discrete-time closed-loop system matrix $(I + A_s)$ will always have an eigenvalue of 1. The consequence of this is that the system will have a linear combination of the pilot voltages that cannot be moved by any control actions, i.e. not all pilot voltages can be fully controlled. In other words, steady state errors are inevitable for the chosen pilot voltages. To fully control the pilot voltages, it is required that the pilot points are selected such that C_c is of full rank.

The secondary level control design is to choose the appropriate gain G_s. The conventional control design neglects the effect of neighboring regions, i.e. it assumes $f[k] = 0$. Under this simplification, the model for pilot voltage dynamics becomes

$$V_c[k+1] - V_c[k] = A_s(V_c[k] - V_c[K]) - D_c d_s[k] \qquad (6.55)$$

where $d_s[k]$ is treated as disturbances to the system. The problem of determining the feedback matrix G_s can be formulated as an optimal control problem with some performance criterion. An alternative way is to specify the desired closed-loop dynamics for the pilot voltages. The choice commonly used in EDF is to specify the closed-loop dynamics so that all pilot load voltages are completely decoupled with each other and exponentially reaching their set values within a specified time constant. This can be easily done by choosing the closed-loop system matrix $(I + A_s)$ to be a fully decoupled diagonal matrix with desired time constant. An example for the choice of time constant typically used in France is three (3) minutes. Equivalently, this is achieved by choosing

$$A_s = \lambda I \qquad (6.56)$$

where λ is a scalar such that the pilot voltages settle to their steady state in the given time. For the specified decoupled dynamics, the time domain response of all pilot voltages will be purely exponential and no overshoot or undershoot will occur. Using the given A_s, one can solve for the gain matrix as

$$G_s = C_c^{-1} \lambda \qquad (6.57)$$

under the assumption that pilot voltages are well chosen such that matrix C_c is nonsingular.

Under the conventional control, the actual dynamics of the pilot voltages become

$$V_c[k+1] - V_c[k] = \lambda(V_c[k] - V_c[K]) + D_c(f[k] - d_s[k]) \qquad (6.58)$$

The flow $f[k]$ is a function of the state variables. Therefore the effective dynamics of the pilot load voltages are not purely exponential. In fact, overshoot or undershoot have been observed in some cases. The numerical example to be given later will illustrate the phenomenon.

6.5.2 Improved secondary-level control

We propose in this section possible ways to improve secondary-level voltage control by taking into consideration the effect of interconnections while preserving its decentralized nature. The proposed control laws will be such that they cancel out the effect of interactions based on additional feedback signals that use the reactive power flow measurements.

It is clear from (6.52) or (6.54) that tie-line flows viewed as an independent external input to the system affect the voltage dynamics. The conventional design of secondary control, i.e., the design of G_s, is typically done neglecting the interconnections with neighboring regions, because of the large scale of the system and the desire to maintain the decentralized nature of secondary-level control. The "optimal" control designed this way will in general not be optimal when implemented on an actual system where interconnections are indeed present. To fully compensate for the effect of interconnections, we propose a new control feedback law in the form

$$u_s[k] = G_s(V_c[k] - V_c[K]) + Hf[k] \qquad (6.59)$$

where the first term is the same as for conventional secondary-level control. The additional term $Hf[k]$ is to cancel the tie-line flows on the dynamics of the pilot voltages. It will be shown that complete cancellation of the tie-line flows for the dynamics of the pilot load voltages is possible with an appropriate choice of the matrix H. Substituting this improved control law into (6.50) leads to

$$V_c[k+1] - V_c[k] = A_s(V_c[k] - V_c[K]) + (C_cH + D_c)f[k] - D_cd_s[k] \qquad (6.60)$$

It is clear that, when C_c is invertible, the tie-line flows can be fully eliminated in the pilot load voltage dynamics by choosing

$$H = -C_c^{-1}D_c \qquad (6.61)$$

With this choice of H, (6.60) reads

$$V_c[k+1] - V_c[k] = A_s(V_c[k] - V_c[K]) - D_cd_s[k] \qquad (6.62)$$

with no flows entering into the equation. In other words, the region under study looks as if it were completely isolated from the rest of the system, as far as pilot load voltage dynamics are concerned.

It is noted that because of the controllability issue of Proposition 6.1, tie-line flows can be fully canceled only for as many as m (the number of active controls) load voltages. Since we choose m pilot points, flows can be canceled for all pilot load voltages.

Note that the control scheme presented here is fully decentralized, assuming the tie-line flows are locally measurable from each region. No detailed information about neighboring regions is needed; only tie-line flows, which aggregate the net effect of detailed dynamics of the neighboring regions, are required. It is not an unrealistic assumption that reactive-power tie-line flows are locally measurable.

6.5.3 The 9-bus example

Let us now illustrate the results by the small 9-bus example network given in Fig. 6.3. Fig. 6.4-(a) shows the pilot load voltage responses using the conventional feedback law given in (6.51). It is seen that because of tie-line interactions, overshoot occurs and settling time is longer than supposed. Fig. 6.4-(b) shows the pilot load voltage responses with the improved feedback control given in (6.59).

It is clear that the additional term $Hf[k]$ improves the responses, both eliminating the overshoot and ensuring the prompt settling. This improvement is expected to be significant when tie lines are strong and meshed.

Figure 6.5 shows the comparison between the non-pilot load voltage responses using the conventional and improved feedback controls. Again improvement is appreciable.

These figures also show that no oscillatory modes exist in the time domain responses of the load voltages, for either conventional secondary control or improved secondary control.

6.6 Tertiary coordination

With an increased tendency toward large energy transfers over far distances, the problem of maintaining voltages within acceptable operating specifications has emerged in operating and planning power systems throughout the world. The systemwide voltage coordination problem is viewed by many as a specialized OPF problem. There are some drawbacks to the OPF technique. First of all, it does not offer much engineering insight for interpreting numerical solutions. As a consequence, when there are convergence problems, OPF cannot explain if the cause is the lack of solutions or the particular numerical method involved. Also, OPF does not offer much opportunity for handling the specific problems of individual regions. Another drawback is that a large amount of data for almost all variables of the entire system is needed in order to carry out the optimization. Therefore, state estimations are necessary for those unmeasured states. Errors in the state estimations can cause problems for the algorithm.

The main purpose of tertiary-level voltage control is to update set values for reactive power tie-line flows $F[K]$, $K = 0, 1, \ldots$ on the tertiary-level time scale in order to optimize systemwide performance for the anticipated base load $Q_L[K]$, $K = 0, 1, \ldots$. This could be done on an hourly basis, if not more often, in accordance with the statistical information on base load. The actual setting of tie-line flows is achieved by changing the settings of secondary voltage controllers $V_G^{set}[K]$. Because this is done so infrequently, it could involve recomputing the basic matrices around a new operating point for the anticipated load over the time horizon T_t.

The idea behind the coordination of regional controllers is to establish feasible schemes for maintaining voltages throughout the interconnected system within the prespecified limits, subject to available reactive power resources. Although in this text only reactive power reserves of generators are of direct interest, the coordination is directly applicable to all other sources of reactive energy that have primary controls responding to local voltages, such as static Var compensators and under load tap changing transformers.

The determination of the optimal set values for the pilot load voltages is formulated as an optimization problem. However, the notion of *optimal* voltage profiles remains an open research question, even for the simplest possible network with one generator supplying power to a single load [62]. This is because the optimal operation of the network has not yet been rigorously defined. The main performance candidates are concerned with

- System reactive reserves.
- Transmission losses.
- Voltage proximity to the prespecified limits.
- Flow scheduling.

We recognize that some performance criteria may be more relevant for normal operating conditions and others for emergency conditions. Therefore, coordination strategies may depend on the system operating mode. In this sense, there will be a certain degree of adaptation to the operating mode. Conventional thinking is that under normal operation one wishes to minimize transmission loss, assuming that the system is well within the reactive reserve and voltage limits.

The optimization problem for the optimal set values of the pilot load voltages can be formulated in three basic topologies, fully centralized, fully decentralized and combined centralized/decentralized, as described in Chapter 4.

6.6.1 Performance criteria

Assuming secondary-level control is carried out properly, the objective of the tertiary-level control is to determine the values for the pilot load voltages, or equivalently, the set values for the generator voltages, over the tertiary-level time scale T_t, so that the global system as a whole operates optimally according to a certain performance criterion. In this section, we discuss some general aspects of the performance criteria to be used for the optimization process.

Since the system is composed of three major components—the generators, the transmission network, and the loads—the overall performance criterion can be written as

$$J = J_{gen} + J_{net} + J_{lod} \tag{6.63}$$

where J_{gen}, J_{net} and J_{lod} are the performance criteria corresponding to each of the three major components of the power system. Specifically, using the above performance criterion, we can achieve the following:

- *Generation alignment* to nearly equalize the ratios of actual generation to the maximum capacity of all or part of the generators.
- *Flow scheduling* to schedule tie-line flows among the interconnected regions so that the global system operates in a coordinated fashion.
- *Security enhancement* to ensure that the generators stay within their limits as much as possible.
- *Loss minimization* to reduce the losses on the transmission network.

For the generators, we need to deal with both the reactive power generation and the terminal voltages. Therefore, the choice for the generator performance criterion can be further decomposed as

$$J_{gen} = J_Q + J_V \tag{6.64}$$

with J_Q pertaining to reactive power generation, and J_V to the voltage limit problem. The term J_Q is introduced to ensure that the reactive power generation remains within physically permissible limits. One simple quadratic form, for example, can be

$$J_Q = (Q_G[K] - Q_G^{nom})^T W_Q (Q_G[K] - Q_G^{nom}) \tag{6.65}$$

where Q_G^{nom} is the desired nominal point inside the limit band of the reactive power generation, and the weighting matrix $W_Q = W_Q^T \geq 0$. The term J_V is primarily to ensure that the generator terminal voltages stay within the allowable bounds. The simple quadratic form for J_V is expressed as

$$J_V = (V_G[K] - V_G^{nom})^T W_V (V_G[K] - V_G^{nom}) \tag{6.66}$$

where V_G^{nom} is the desired nominal point inside the limit band of the generator terminal voltages, and the weighting matrix $W_V = W_V^T \geq 0$. This kind of performance criterion tends to keep the generator generation outputs and terminal voltages close to their desired nominal values, if heavy weights are given to these terms. The justification for this type of performance criterion is that it can eliminate the situation of some generators hitting their physical operating limits under heavy loading conditions.

Similarly, for the transmission network, we can decompose the performance criterion into a term involving the total losses on the transmission network and a term involving rescheduling the tie-line flows. For the loads, the primary concern is also for the critical pilot node load voltages to stay within acceptable bounds. An expression similar to (6.66) can be written for the load voltages:

$$J_{lod} = (V_c[K] - V_c^{nom})^T W_c (V_c[K] - V_c^{nom}) \tag{6.67}$$

where V_c^{nom} is the desired point inside the limit band of the load voltages, and the weighting matrix $W_c = W_c^T > 0$. This performance criterion tends to keep the load voltages close to the desired point V_c^{nom}.

In the process of solving the optimal control problem given in (6.63), constraints among the tie-line flows, unit generation outputs, losses, and the set values for the pilot load voltages on the tertiary-level time scale are needed. These constraints for the time K involve the values of these quantities at the previous time $(K-1)$. As a consequence, the optimal solution for the time K involves quantities at the previous time $(K-1)$. Therefore, the pilot load voltage settings as a result of the optimal control problem form another discrete-time sequence on the very slow time scale T_t. The basic requirement for the optimization process is that it must guarantee the stability of this discrete event process.

It is emphasized that this process of tertiary-level control involves only information about the generators participating in secondary-level control and about the pilot point loads, plus the tie-line flows if they are to be rescheduled. The amount of data and computation involved is much less than that needed for full-scale optimal power flow calculation, where information about all loads is necessary. As an example, two regions of the French network have 259 buses, while the number of pilot points for the two regions is only 9. Because of this reduced information needed, it is possible that this tertiary-level control scheme can be implemented on line as closed-loop control.

6.7 New tertiary level-aggregate models

Using the structural modeling approach discussed above, one can derive the relation between the critical pilot load voltages and the generator voltages, the critical pilot load voltages and the reactive power outputs of generators, the flows and the interaction variables, and the flows and the output variables, on the tertiary level time scale T_t. These relations serve as constraints to the optimization problem to determine the optimal set values for the output variables—the critical pilot load voltages. Since these relations are derived for an administrative region, they explicitly involve the tie-line flows into the region from the rest of the system.

Pilot load voltages and generator voltages

Let us consider a single region within an interconnected system. The relation between the load and generator voltages was introduced in (6.19) as

$$\dot{V}_L = C_V \dot{V}_G + J_{LL}^{-1}(\dot{F}_L - \dot{Q}_L) \tag{6.68}$$

To derive the relation between the critical pilot load voltages and the generator voltages on the tertiary-level time scale T_t, let us integrate this equation from KT_t to $(K+1)T_t$. This simply leads to

$$\begin{aligned} V_L[K+1] - V_L[K] &= C_V(V_G[K+1] - V_G[K]) + \\ &\quad D(F_L[K+1] - F_L[K]) - Dd_t[K] \end{aligned} \tag{6.69}$$

where $D \triangleq J_{LL}^{-1}$, and $d_t[K] \triangleq Q_L[K+1] - Q_L[K]$. The critical pilot load voltages are defined as

$$V_c \triangleq C_s V_L \tag{6.70}$$

Multiplying (6.69) by C_s yields

$$\begin{aligned} V_c[K+1] - V_c[K] &= C_c(V_G[K+1] - V_G[K]) + D_c(F_L[K+1] \\ &\quad - F_L[K]) - D_c d_t[K] \end{aligned} \tag{6.71}$$

where $C_c \triangleq C_s C_V$ is the sensitivity matrix of the critical pilot load voltages relative to the control, and $D_c = C_s D$. Following the discussion in Chapter 4, one should note that C_c is chosen to be a square matrix and nonsingular. Otherwise, not all pilot load voltages can be fully controlled. Under this condition, one obtains from (6.71)

$$\begin{aligned} V_G[K+1] - V_G[K] &= L_V(V_c[K+1] - V_c[K]) - \\ &\quad L_{Vd}(F_L[K+1] - F_L[K]) + L_{Vd} d_t[K] \end{aligned} \tag{6.72}$$

where $L_V \triangleq C_c^{-1}$ and $L_{Vd} \triangleq L_V D_c$. This relation determines the change in generator voltages for any change in the pilot load voltages and the tie-line flows under the given loading condition. Let us rewrite this in a more compact form as

$$V_G[K+1] = L_V V_c[K+1] - L_{Vd} F_L[K+1] + R_V[K] \tag{6.73}$$

where

$$R_V[K] = V_G[K] - L_V V_c[K] + L_{Vd} F_L[K] + L_{Vd} d_t[K] \tag{6.74}$$

Or equivalently, with a time delay,

$$V_G[K] = L_V V_c[K] - L_{Vd}F_L[K] + R_V[K-1] \tag{6.75}$$

This relation will be used by the decentralized regional tertiary-level control to determine the regional pilot voltages.

Pilot load voltages and reactive generation

Another important quantity of concern to tertiary-level control is the reactive power generation of the generating units, because the reactive power generation must match the reactive loads and the losses on the transmission lines. Here we derive the relation between the changes in generator reactive outputs and the changes in pilot load voltages. This relation is important because one needs to be concerned with the reactive power generation when deciding on the desired pilot load voltages. The fundamental relation between the reactive power generation and the generator voltages was derived in (6.23) as

$$\dot{Q}_G = K_Q \dot{V}_G - \dot{F}_e + D_Q \dot{Q}_L \tag{6.76}$$

where $K_Q = J_{GG} + J_{GL}C_V$, and $D_Q = -J_{GL}J_{LL}^{-1}$. With the same derivations as in the previous section, by integrating this on the tertiary-level time scale T_t, we have

$$\begin{aligned}Q_G[K+1] - Q_G[K] &= K_Q(V_G[K+1] - V_G[K]) - (F_e[K+1] - F_e[K]) \\ &\quad + D_Q d_t[K]\end{aligned} \tag{6.77}$$

This, combined with (6.72), gives us

$$\begin{aligned}Q_G[K+1] - Q_G[K] &= L_Q(V_c[K+1] - V_c[K]) - (F_Q[K+1] - \\ &\quad F_Q[K]) + L_{Qd}d_t[K]\end{aligned} \tag{6.78}$$

where $L_Q \triangleq K_Q L_V$, $F_Q \triangleq F_e + K_Q L_{Vd} F_L$, and $L_{Qd} \triangleq D_Q + K_Q L_{Vd}$. Eq. (6.78) defines the relation between the changes in reactive power generation and the changes in pilot load voltages, given the tie-line flow and load changes. Again, we can rewrite this into

$$Q_G[K+1] = L_Q V_c[K+1] - F_Q[K+1] + R_Q[K] \tag{6.79}$$

where

$$R_Q[K] = Q_G[K] - L_Q V_c[K] + F_Q[K] + L_{Qd}d_t[K] \tag{6.80}$$

Or equivalently, with a time delay,

$$Q_G[K] = L_Q V_c[K] - F_Q[K] + R_Q[K-1] \tag{6.81}$$

This equation will also serve as a constraint for the regional tertiary-level optimization process.

Flows and the interaction variables

We have derived the secondary-level interaction dynamic model in (6.41) as

$$z[k+1] - z[k] = TD_s(f[k] - d_s[k]) \tag{6.82}$$

or

$$z[k+1] - z[k] = S(f[k] - d_s[k]) \tag{6.83}$$

where

$$S \triangleq TD_s \tag{6.84}$$

Evaluating this on the tertiary-level time scale T_t yields

$$z[K+1] - z[K] = S(F_L[K+1] - F_L[K]) - S(d[K+1] - d[K]) \tag{6.85}$$

Define $d_t[K] \triangleq d[K+1] - d[K]$ as the tertiary-level disturbance. Eq. (6.85) becomes

$$z[K+1] - z[K] = S(F_L[K+1] - F_L[K]) - Sd_t[K] \tag{6.86}$$

Matrix S as defined in (6.84) has a special structure and can be constructed using inspection. Recall the definition of $C_V = -J_{LL}^{-1} J_{LG} = -D_s J_{LG}$ and the condition $TC_V = 0$ in (6.40). One simply has

$$SJ_{LG} = TD_s J_{LG} = -TC_V \tag{6.87}$$

Therefore we derive the important equation for the S matrix as

$$SJ_{LG} = 0 \tag{6.88}$$

This gives a simple method to construct the S matrix. Suppose that there are n load buses and m generator buses in the region under consideration, and assume that $n > m$. Then there are $n - m$ independent solutions for S, or S has $n-m$ independent rows. If load bus i is not connected to a generator bus, then the corresponding ith row of J_{LG} is all zero, because J_{LG} is the connection matrix between the load buses and the generator buses. In this case, vector

$$v = \begin{bmatrix} 0 & \cdots & 0 & 1 & 0 & \cdots & 0 \end{bmatrix} \tag{6.89}$$
$$\uparrow\ i^{th}\ element$$

will satisfy $vJ_{LG} = 0$ because the ith row of J_{LG} is all zero. In other words, matrix S will have v as one of its rows. Therefore, for all load buses not connected to generator buses, we construct S by selecting 1 at the corresponding locations, and 0 elsewhere. If the number of load buses not connected to generator buses is equal to $n - m$, then we have found all $n - m$ independent solutions for S. If this number is less than $n - m$, then there are more independent solutions to be determined. In this case, one needs to solve (6.88) to get all the independent solutions.

Because of the special structure of the S matrix, we see from (6.86) that the interaction variables are just the tie-line flows into all the load buses that are not connected to generator buses. This is extremely important for tertiary-level coordination.

6.7.1 Centralized aggregate models

In this section, the global interconnected system is considered as one single region. Since this big single region is an isolated system, there are no tie-line flows into the system. All the previous derivations carry over to the global interconnected system, except all tie-line flow terms drop out. This is because there is no restriction on how to choose the region, and all results apply to the global interconnected system as one single region.

Pilot load voltages and generator voltages

We use bold face letters to represent any variable associated with the global interconnected system. For example, we define $\boldsymbol{V}_G$, $\boldsymbol{Q}_G$ to represent the generator voltages and reactive power generation of the global system, and $\boldsymbol{V}_c$ to represent the critical pilot load voltages of the global system. The relation between the pilot load voltages and the generator voltages is in the same form as (6.75):

$$\boldsymbol{V}_G[K] = \boldsymbol{L}_V \boldsymbol{V}_c[K] + \boldsymbol{R}_V[K-1] \tag{6.90}$$

where $\boldsymbol{L}_V \triangleq \boldsymbol{C}_c^{-1}$ and $\boldsymbol{L}_{Vd} \triangleq \boldsymbol{L}_V \boldsymbol{D}_c$. Matrix $\boldsymbol{C}_c$ is the pilot voltage sensitivity matrix for the global system, and

$$\boldsymbol{R}_V[K-1] = \boldsymbol{V}_G[K-1] - \boldsymbol{L}_V \boldsymbol{V}_c[K-1] + \boldsymbol{L}_{Vd}\boldsymbol{d}_t[K-1] \tag{6.91}$$

Comparing with (6.75), we see that all the flow terms disappear here because the global system is assumed to be an isolated one, and there is no flow for the isolated system.

Pilot load voltages and reactive power generation

As with the generator voltage case, we can derive the tertiary-level aggregate model for the reactive power generation of the global system. Again, we carry the results for a single region case. The final relation needed for tertiary-level control is derived from (6.81) without all the flow terms:

$$Q_G[K] = L_Q V_c[K] + R_Q[K-1] \tag{6.92}$$

where $L_Q \triangleq K_Q L_V$, and

$$R_Q[K-1] = Q_G[K-1] - L_Q V_c[K-1] + L_{Qd} d_t[K-1] \tag{6.93}$$

All other matrices are defined in a similar way as in the decentralized case, except they are now defined for the global system.

Flows and the pilot load voltages

Here we study the relation between the internal flows among the regions within the global system and the pilot load voltages. This relation can be easily obtained from secondary-level quasi-static models as

$$S(F[K+1] - F[K]) = L(y_s[K+1] - y_s[K]) + L_d d_t[K] \tag{6.94}$$

where $S = TD_{LL}$ and $L = TC_V(TC_c)^{-1}$.

It is intriguing to recognize that $z[k]$ can be interpreted as the area load excess (ALE), introduced in [4] as a better alternative to the area control error (ACE) signal. This simply follows from the Kirchoff's current law for each cutset separating an area from its neighboring subsystems. While the ALE concept was introduced in [4] as a heuristic measure of the most meaningful signal for preserving the area control principle, we provide in this chapter its accurate derivation, which also accounts for transmission losses. The basic difference from the previous literature, including [3] and [4], is that it is not necessary to experiment with the best weighting coefficients when designing secondary-level and tertiary-level controllers. They simply result from general optimization methods for chosen performance criteria.

6.7.2 The 9-bus example

Let us illustrate the general model by considering the 9-bus example given in Fig. 6.3. For the numbers given in Table 6.1, we can calculate

$$C_c = \begin{bmatrix} 1 & 0 & 0 \\ 0 & .55 & .45 \\ 0 & .37 & .63 \end{bmatrix} \tag{6.95}$$

$$L_V = C_c^{-1} = \begin{bmatrix} 1 & 0 & 0 \\ 0 & 3.43 & -2.43 \\ 0 & -2.02 & 3.02 \end{bmatrix} \tag{6.96}$$

Matrix $\boldsymbol{K_Q}$ is given by

$$\boldsymbol{K_Q} = \begin{bmatrix} 0 & 0 & 0 \\ 0 & 3.71 & -3.71 \\ 0 & -3.71 & 3.71 \end{bmatrix} \qquad (6.97)$$

and we can calculate

$$\boldsymbol{L_Q} = \boldsymbol{K_Q L_V} = \begin{bmatrix} 0 & 0 & 0 \\ 0 & 20.24 & -20.24 \\ 0 & -20.24 & 20.24 \end{bmatrix} \qquad (6.98)$$

The $\boldsymbol{D_{LL}}$ matrix is

$$\boldsymbol{D_{LL}} = \begin{bmatrix} -7.69 & 0 & 0 & 1 & 0 & 0 \\ 0 & -52.75 & 15.67 & 21.55 & 0 & 7.19 \\ 0 & 15.67 & -38.16 & 0 & 0 & 12.49 \\ 0 & 21.55 & 0 & -21.55 & 0 & 0 \\ 0 & 0 & 0 & 0 & -9.89.8 \\ 0 & 7.19 & 12.49 & 0 & 9.8 & -29.48 \end{bmatrix} \qquad (6.99)$$

With the $\boldsymbol{T}$ matrix given in (6.45), we obtain

$$\boldsymbol{S} = \boldsymbol{T D_{LL}} = \begin{bmatrix} 0 & 0 & 0 & 1 & 0 & 0 \\ 0 & 0 & 0 & 0 & 1 & 0 \\ 0 & 0 & 0 & 0 & 0 & 1 \end{bmatrix} \qquad (6.100)$$

$$\boldsymbol{L} = \begin{bmatrix} -4.06 & 4.06 & 0 \\ -3.38 & 1.23 & 2.15 \\ 0 & 0 & 0 \end{bmatrix} \qquad (6.101)$$

and

$$\boldsymbol{L_d} = (\boldsymbol{T} - \boldsymbol{LC})\boldsymbol{D_s} = \begin{bmatrix} -.53 & .27 & .18 & 1.27 & .21 & .21 \\ -.44 & .18 & .19 & .18 & 1.19 & .19 \\ 0 & 0 & 0 & 0 & 0 & 1 \end{bmatrix} \qquad (6.102)$$

The physical meaning of the matrix $\boldsymbol{S}$ is that only the tie-line flows going into the load buses are the interaction variables. The fact that all elements of the last row of matrix $\boldsymbol{L}$ are zero means that the third interaction variable $z_3 = 7.19(x_2 - x_6) + 12.49(x_3 - x_6) + 9.8(x_5 - x_6)$ as calculated in (6.46) will remain constant even when the tie lines are connected, if there are no load variations. This physically makes sense because this interaction variable is always equal to the load at bus #6, as can be seen from Fig. 6.3, no matter what the generator voltages or the pilot load voltages are. This is clearly demonstrated by the last row of matrix $\boldsymbol{L_d}$.

6.7.3 Fully centralized optimization

In this section, we formulate the tertiary-level optimization for the fully centralized case, i.e., the coordination tasks are carried out by a tertiary-level control center. This control center is assumed to have all information about the interconnected regions needed to solve the optimization problem.

Consider an interconnected system consisting of R regions. Let $\mathbf{V}_G$, $\mathbf{Q}_G$, and $\mathbf{V}_c$ represent the generator voltages, the reactive power generation, and the critical pilot load voltages of the global system. Let the performance criterion for the global system in the interval $[KT_t, (K+1)T_t]$ be given by

$$J[K] = J(\mathbf{V}_G[K], \mathbf{Q}_G[K], \mathbf{V}_c[K]) \tag{6.103}$$

Let us take, as an example, the quadratic form

$$\begin{aligned} J[K] &= (\mathbf{V}_G[K] - \mathbf{V}_G^{nom})^T \mathbf{W}_V (\mathbf{V}_G[K] - \mathbf{V}_G^{nom}) \\ &+ (\mathbf{Q}_G[K] - \mathbf{Q}_G^{nom})^T \mathbf{W}_Q (\mathbf{Q}_G[K] - \mathbf{Q}_G^{nom}) \\ &+ (\mathbf{V}_c[K] - \mathbf{V}_c^{nom})^T \mathbf{W}_c (\mathbf{V}_c[K] - \mathbf{V}_c^{nom}) \end{aligned} \tag{6.104}$$

where $\mathbf{V}_G^{nom}$, $\mathbf{Q}_G^{nom}$, and $\mathbf{V}_c^{nom}$ are the desired nominal values for the generator voltages, reactive power generation, and pilot load voltages, and $\mathbf{W}_V$, $\mathbf{W}_Q$, and $\mathbf{W}_c$ are the relative weighing matrices for the corresponding terms. The purpose of the optimization process is to determine the optimal setting for the pilot load voltages $\mathbf{V}_c[K]$.

Note that the generator voltages $\mathbf{V}_G$ and reactive power generation $\mathbf{Q}_G$ are related to the pilot load voltage settings through the aggregate models presented in the previous section. These relations serve as equality constraints for the optimization criterion proposed in (6.104). This optimization problem, together with these constraints, can be explicitly solved and the analytic solution can be obtained. For that purpose, let us first prove the following optimization result:

- **Proposition 6.2. (Optimal solution):** The analytic solution to the constrained optimization problem,

$$\begin{cases} J = (x - x^{nom})^T Q (x - x^{nom}) + (y - y^{nom})^T R (y - y^{nom}) \\ x = Ly + x_0 \end{cases} \tag{6.105}$$

  is explicitly given by

$$\begin{cases} x = x^{nom} + X(x^{nom} - Ly^{nom} - x_0) \\ y = y^{nom} + Y(x^{nom} - Ly^{nom} - x_0) \end{cases} \tag{6.106}$$

  where

$$\begin{cases} X = LY - I \\ Y = (L^T Q L + R)^{-1} L^T Q \end{cases} \tag{6.107}$$

Chapter 6

with I being the identity matrix of the same dimension as Q.

Note that if $x^{nom} = Ly^{nom} + x_0$, i.e., the nominal values satisfy the constraint between x and y, then the optimal solution is given by $x = x^{nom}$ and $y = y^{nom}$. This is straightforward. In this case, the optimal cost is calculated to be $J = 0$.

Proof: Let us convert the constrained optimization problem into an unconstrained one by constituting a Lagrangian

$$\mathcal{L} = (x-x^{nom})^T Q(x-x^{nom}) + (y-y^{nom})^T R(y-y^{nom}) - \lambda^T(x-Ly-x_0) \quad (6.108)$$

Using vector differentiation results, we derive

$$\frac{\partial \mathcal{L}}{\partial x} = 2Q(x - x^{nom}) - \lambda \quad (6.109)$$

$$\frac{\partial \mathcal{L}}{\partial y} = 2R(y - y^{nom}) + L^T \lambda \quad (6.110)$$

By setting the two partial derivatives to zero, we obtain

$$R(y - y^{nom}) + L^T Q(x - x^{nom}) = 0 \quad (6.111)$$

Using the constraint $x = Ly + x_0$, we get

$$R(y - y^{nom}) + L^T Q(Ly + x_0 - x^{nom}) = 0 \quad (6.112)$$

or

$$(L^T QL + R)(y - y^{nom}) = L^T Q(x^{nom} - Ly^{nom} - x_0) \quad (6.113)$$

which leads to the conclusion in the proposition. Having obtained the optimal solution for y, one can calculate the corresponding optimal solution for x from the constraint $x = Ly + x_0$. That simply yields the result stated in the proposition.

Now we can use Proposition 6.2 to solve the optimization problem posed in (6.104). To put the performance criterion into the form given in the proposition, let us define the vector of generator quantities as

$$\chi \triangleq \begin{bmatrix} V_G \\ Q_G \end{bmatrix} \quad (6.114)$$

and the weighting matrix for the generator quantities as

$$\mathcal{Q} \triangleq \text{BlockDiag}(W_V, W_Q) \quad (6.115)$$

The performance criterion in (6.104) can be rewritten as

$$J[K] = (x[K] - x^{nom})^T Q(x[K] - x^{nom})$$
$$+ (V_c[K] - V_c^{nom})^T W_c(V_c[K] - V_c^{nom}) \qquad (6.116)$$

The generator voltages and pilot load voltages satisfy the constraint as derived in (6.90):

$$V_G[K] = L_V V_c[K] + R_V[K-1] \qquad (6.117)$$

The reactive generation and pilot load voltages satisfy the constraint as given in (6.92):

$$Q_G[K] = L_Q V_c[K] + R_Q[K-1] \qquad (6.118)$$

Let us further rewrite these constraints as a single constraint in the form

$$x[K] = \mathcal{L} V_c[K] + \mathcal{R}[K-1] \qquad (6.119)$$

where

$$\mathcal{L} \triangleq \begin{bmatrix} L_V \\ L_Q \end{bmatrix} \quad \text{and} \quad \mathcal{R} \triangleq \begin{bmatrix} R_V \\ R_Q \end{bmatrix} \qquad (6.120)$$

Now the problem is in exactly the same format as Proposition 6.2, with x corresponding to x and $V_c[K]$ corresponding to y of the proposition. The optimal solution for the pilot load voltage settings are simply given by

$$V_c[K] = V_c^{nom} + \mathcal{Y}(x^{nom} - \mathcal{L} V_c^{nom} - \mathcal{R}[K-1]) \qquad (6.121)$$

where

$$\mathcal{Y} = (\mathcal{L}^T Q \mathcal{L} + W_c)^{-1} \mathcal{L}^T Q \qquad (6.122)$$

Corresponding to this optimal solution for the pilot voltage settings, the generator voltages and the reactive power generation outputs are given by

$$x[K] = x^{nom} + \mathcal{X}(x^{nom} - \mathcal{L} V_c^{nom} - \mathcal{R}[K-1]) \qquad (6.123)$$

where

$$\mathcal{X} = \mathcal{L}\mathcal{Y} - \mathcal{I} \qquad (6.124)$$

Note that in the process of solving the optimality problem, inequality constraints such as voltage or reactive generation limits are not explicitly taken into account, for the purpose of simplicity. In real situations, these physical limits must be checked to ensure that all generators operate within permissible ranges.

The obtained optimal solution forms a control-driven discrete event process on the tertiary-level time scale, driven by the load variations $d_t[K]$ through $\mathcal{R}[K-1]$. As a result, the optimal solution will be a function of the load variations so that the global system is kept optimal as loading varies.

6.7.4 Fully decentralized optimization

Here we study tertiary-level level control in the fully decentralized case, i.e., each region within the interconnected system optimizes its own performance criterion, while assuming no structural information about the rest of the system. In the regional optimization process, each region measures the tie-line flows and uses the measurement to determine its optimal settings.

Consider a single region within an interconnected system consisting of R regions. Let V_G, Q_G, and V_c represent the generator voltages, reactive power generation, and critical pilot load voltages of the region under consideration. As with fully centralized optimization, the performance criterion for the region in the interval $[KT_t, (K+1)T_t]$ can be written in the form

$$J[K] = J(V_G[K], Q_G[K], V_c[K]) \tag{6.125}$$

Let us also take the quadratic form

$$\begin{aligned} J[K] &= (V_G[K] - V_G^{nom})^T W_V (V_G[K] - V_G^{nom}) \\ &+ (Q_G[K] - Q_G^{nom})^T W_Q (Q_G[K] - Q_G^{nom}) \\ &+ (V_c[K] - V_c^{nom})^T W_c (V_c[K] - V_c^{nom}) \end{aligned} \tag{6.126}$$

Again, V_G^{nom}, Q_G^{nom}, and V_c^{nom} are the desired nominal values for the generator voltages, reactive power generation, and pilot load voltages of the particular region, W_V, W_Q, and W_c are the relative weighting matrices for the corresponding terms. The purpose of the optimization process is also to determine the optimal setting for the pilot voltages $V_c[K]$, as in the fully centralized optimization case.

The constraints between $V_G[K]$ and $V_c[K]$, $Q_G[K]$ and $V_c[K]$ are as given in (6.75) and (6.81). For convenience, let us repeat them:

$$V_G[K] = L_V V_c[K] - L_{Vd} F_L[K] + R_V[K-1] \tag{6.127}$$

$$Q_G[K] = L_Q V_c[K] - F_Q[K] + R_Q[K-1] \tag{6.128}$$

where $R_V[K-1]$ and $R_Q[K-1]$ are defined as

$$\begin{aligned} R_V[K-1] &= V_G[K-1] - L_V V_c[K-1] + L_{Vd} F_L[K-1] + \\ &\quad L_{Vd} d_t[K-1] \end{aligned} \tag{6.129}$$

$$R_Q[K-1] = Q_G[K-1] - L_Q V_c[K-1] + F_Q[K-1] + L_{Qd} d_t[K-1] \tag{6.130}$$

As with fully centralized optimization case, let us convert the problem into the form in Proposition 6.2. in order to utilize the proposition. Define again the vector of generator quantities for the particular region under study

$$x \triangleq \begin{bmatrix} V_G \\ Q_G \end{bmatrix} \tag{6.131}$$

and its weighting matrix

$$Q \triangleq \text{BlockDiag}(W_V, W_Q) \tag{6.132}$$

The performance criterion in (6.104) can now be rewritten as

$$\begin{aligned}J[K] &= (x[K] - x^{nom})^T Q(x[K] - x^{nom}) \\ &\quad + (V_c[K] - V_c^{nom})^T W_c(V_c[K] - V_c^{nom})\end{aligned} \tag{6.133}$$

The two constraints (6.127) and (6.128) can be put together into

$$x[K] = \mathcal{L}V_c[K] - \mathcal{F}[K] + \mathcal{R}[K-1] \tag{6.134}$$

where

$$\mathcal{L} \triangleq \begin{bmatrix} L_V \\ L_Q \end{bmatrix}, \quad \mathcal{F} \triangleq \begin{bmatrix} L_{Vd} F_L \\ F_Q \end{bmatrix}, \quad \text{and } \mathcal{R} \triangleq \begin{bmatrix} R_V \\ R_Q \end{bmatrix} \tag{6.135}$$

In the optimization process, the tie-line flows $\mathcal{F}[K]$ will be taken as a measured constant. By Proposition 6.2, the resulting optimal solution is given as

$$V_c[K] = V_c^{nom} + \mathcal{Y}(x^{nom} - \mathcal{L}V_c^{nom} + \mathcal{F}[K] - \mathcal{R}[K-1]) \tag{6.136}$$

where

$$\mathcal{Y} = (\mathcal{L}^T Q \mathcal{L} + W_c)^{-1} \mathcal{L}^T Q \tag{6.137}$$

Corresponding to this optimal solution for the pilot load voltage settings, the generator voltages and the reactive power generation outputs are given by

$$x[K] = x^{nom} + \mathcal{X}(x^{nom} - \mathcal{L}V_c^{nom} + \mathcal{F}[K] - \mathcal{R}[K-1]) \tag{6.138}$$

where

$$\mathcal{X} = \mathcal{L}\mathcal{Y} - \mathcal{I} \tag{6.139}$$

All individual regions measure the tie-line flows to determine their optimal pilot voltage settings.

6.7.5 Combined centralized/decentralized optimization

The fully centralized and fully decentralized approaches discussed above represent two extremes: one requires full information about the global system and the other neglects the structural properties of the rest of the system. An approach that lies between the two is the combined centralized/decentralized optimization method, in which one assumes partial information about the rest of the system. The partial information is defined by the aggregate models discussed previously. This approach falls into the category of game theory.

Chapter 6   153

Game formulation for voltage control

In power systems, the decision variables are the generator voltage settings, or equivalently, the critical pilot load voltage settings. For a specific region, the performance criterion usually involves quantities like the transmission losses, reactive power generation, generator voltages, and critical pilot point loads. The decision variables of other regions do not directly enter the expression of the performance criterion for this region. In other words, the competition is not in a standard game theoretical setting.

To see this, we consider again an interconnected system consisting of R regions. Let the performance criterion of any single region be given by

$$J[K] = J(V_G[K], Q_G[K], V_c[K]) \tag{6.140}$$

Note that all variables in this expression are associated with this particular region only; no variables associated with other regions directly enter the function. The coupling among the regions occurs only when the constraints among $V_G[K]$, $Q_G[K]$, and $V_c[K]$ are introduced. These constraints involve the inter-regional tie-line flows, which couple all interconnected regions.

To see this, let us rewrite (6.140) as

$$J[K] = J(x[K], V_c[K]) \tag{6.141}$$

with x being defined as in (6.131). The constraint between x and V_c was derived in (6.134) as

$$x[K] = \mathcal{L}V_c[K] - \mathcal{F}[K] + \mathcal{R}[K-1] \tag{6.142}$$

where all quantities are as defined previously. In this equation, the tie-line flow $\mathcal{F}[K]$ acts to couple different regions, because it is a function of the generator or load voltages of all the involved regions. This function can be expressed as

$$\begin{aligned}\mathcal{F}[K] &= \mathcal{N}_G V_G[K] + \mathcal{N}_c V_c[K] \\ &= \mathcal{N}_G^I V_G^I[K] + \cdots + \mathcal{N}_G^R V_G^R[K] \\ &\quad + \mathcal{N}_c^I V_c^I[K] + \cdots + \mathcal{N}_c^R V_c^R[K]\end{aligned} \tag{6.143}$$

where V_G and V_c are the generator voltages and critical pilot load voltages of the global system consisting of all regions involved, and matrices $\mathcal{N}_G$ and $\mathcal{N}_c$ are related to the strength of interconnections. When all interconnections are removed, $\mathcal{N}_G$ and $\mathcal{N}_c$ are both zero.

In light of this constraint, the performance criterion in (6.141) becomes a function of the pilot load voltages, and the generator voltages of all the regions within the interconnected global system. In other words, (6.141) can be represented in the form of

$$J[K] = J(V_G[K], V_c[K]) \tag{6.144}$$

With this performance criterion, we define the reaction curves for this particular region under study as

$$\frac{\partial J}{\partial V_c} = 0 \tag{6.145}$$

The solution corresponding to the intersection of these reaction curves is called the Nash strategy for the power system. This definition is a natural extension of the standard Nash reaction curves.

Let us explicitly find the Nash solution for a quadratic performance criterion

$$\begin{aligned} J[K] &= (x[K] - x^{nom})^T Q(x[K] - x^{nom}) \\ &+ (V_c[K] - V_c^{nom})^T W_c(V_c[K] - V_c^{nom}) \end{aligned} \tag{6.146}$$

To do this, we need the following vector differentiation results:

Vector Differentiation: Let x and y, $y = y(x)$, be column vectors, and let A be a matrix (independent of x) of appropriate dimension such that all operations are meaningful. The following is true:

$$\frac{\partial}{\partial x}\left[x^T A y \right] = A y + \left[A \frac{\partial y}{\partial x} \right]^T x \tag{6.147}$$

$$\frac{\partial}{\partial x}\left[y^T A y \right] = \left[\frac{\partial y}{\partial x} \right]^T (A + A^T) y \tag{6.148}$$

where the matrix

$$\frac{\partial y}{\partial x} \triangleq \left[\frac{\partial y_i}{\partial x_j}\right]_{i,j} \tag{6.149}$$

One can verify these two equations by expanding the expressions and doing element-by-element differentiation. Details are not given here.

Now we are ready to derive the analytic Nash solution for the posed problem. From (6.146), using the vector differentiation results, one has

$$\frac{\partial J}{\partial V_c} = 2\left[\frac{\partial x}{\partial V_c}\right]^T Q(x[K] - x^{nom}) + 2W_c(V_c[K] - V_c^{nom}) = 0 \tag{6.150}$$

Matrix $\dfrac{\partial x}{\partial V_c}$ can be found from (6.142) as

$$\frac{\partial x}{\partial V_c} = \mathcal{L} - \frac{\partial \mathcal{F}}{\partial V_c} \tag{6.151}$$

Using (6.143), one obtains

$$\frac{\partial \mathcal{F}}{\partial V_c} = \mathcal{N}_c + \mathcal{N}_G \frac{\partial V_G}{\partial V_c} = \mathcal{N}_c + \mathcal{N}_G L_V \qquad (6.152)$$

Therefore

$$\frac{\partial x}{\partial V_c} = \mathcal{L} - \mathcal{N}_c - \mathcal{N}_G L_V \qquad (6.153)$$

From this equation, we can rewrite (6.150) as

$$(\mathcal{L} - \mathcal{N}_c - \mathcal{N}_G L_V)^T Q(x[K] - x^{nom}) + W_c(V_c[K] - V_c^{nom}) = 0 \qquad (6.154)$$

Together with the constraint relation (6.142), we obtain

$$V_c[K] = V_c^{nom} + \mathcal{Y}(x^{nom} - \mathcal{L}V_c^{nom} + \mathcal{F}[K] - \mathcal{R}[K-1]) \qquad (6.155)$$

where $\mathcal{Y}$ is defined by

$$\mathcal{Y} = [(\mathcal{L} - \mathcal{N}_c - \mathcal{N}_G L_V)^T Q\mathcal{L} + W_c]^{-1}(\mathcal{L} - \mathcal{N}_c - \mathcal{N}_G L_V)^T Q \qquad (6.156)$$

In this optimization result, the optimal pilot load voltage settings are expressed explicitly in terms of the tie-line flows, which are measured by each individual region. As with the previous cases, we can find the generator voltages and the reactive power generation, corresponding to this optimal solution for the pilot voltage settings, as

$$x[K] = x^{nom} + \mathcal{X}(x^{nom} - \mathcal{L}V_c^{nom} + \mathcal{F}[K] - \mathcal{R}[K-1]) \qquad (6.157)$$

with

$$\mathcal{X} = \mathcal{L}\mathcal{Y} - \mathcal{I} \qquad (6.158)$$

Again, if $x^{nom} = \mathcal{L}V_c^{nom} - \mathcal{F}[K] + \mathcal{R}[K-1]$, i.e. the two nominal values also satisfy the constraint, then the optimal solution is simply the nominal values.

6.7.6 Simulations study of the French power network

In this section, we describe the French power network as an example to illustrate the concepts and results developed in this chapter. The most serious coordination problems are described to us by EDF personnel as occurring for two regions on the EDF network. The entire French power network consists of more than 1,000 nodes and is divided into seven regions. The two most important regions for the study of coordination are the eastern part

Regions	Pilot Nodes	Control Units
III	COULAS71	CRUA5T 1
III	TRI.PS61	TRICAT 1
III	TAVELS71	ARAMOT 1
III	SEPTES61	M.PONT 1
IV	CHAFFS71	BUGEYT 2
IV	P.CORS71	SSAL7T 1
IV	GIVORS61	LOIRET 3
IV	CPNIES61	VAUJH 7
IV	ALBERS71	S.BIH 4

Table 6.2: Pilot nodes and control units of EDF network

and southeastern part of France. To be consistent with the French division, refer to the two regions as region III and region IV. The boundary interconnections are depicted in Fig. 6.6.

These two regions are highly interconnected through two strong tie lines, COULAS71 to CHAFFS71, and COULAS71 to P.CORS71. There are also two weak tie lines connecting the two regions, BOUDES61 to MTPEZS61, and P.BORS61 to PRATCS63. There are in total 205 nodes in these two regions. Region III has 4 pilot nodes, and region IV has 5 pilot nodes. The pilot load voltages and the generating units participating in secondary-level control are listed in Table 6.2.

The practical problems that may arise when regional secondary-level voltage controls are not appropriately coordinated can be classified into dynamic problems and static problems. The dynamic problems are seen, for example, as the overshooting of pilot load voltages because of coupling among the pilot load voltages, or reactive generation outputs of units moving in opposite directions during the transient process. The static problems arise when each region chooses its set points for the pilot load voltages without taking into account the neighboring regions. Such scenarios may result in excessive and useless reactive power exchanges on the tie lines or a large imbalance of reactive generations on the system. In such cases, the ability of the network to handle rapid cascades of endangering events can be greatly reduced.

Here we use the coordination schemes described in the previous sections to address these problems. It will be shown that our theory for tertiary-level coordination successfully solves these problems. The simulation results show that it is feasible and efficient to achieve coordination by directly controlling a few selected pilot load voltages at the subsystem level.

Regions	Control Units	Maximum	Before Control	After Control
III	CRUA5T 1	2111	455	844
III	TRICAT 1	488	198	258
III	ARAMOT 1	674	386	312
III	M.PONT 1	619	283	281
IV	BUGEYT 2	976	965	936
IV	SSAL7T 1	1438	1166	681
IV	LOIRET 3	309	−83	94
IV	VAUJH 7	889	296	425
IV	S.BIH 4	371	249	191

Table 6.3: Maximum and actual generation of EDF network, MVar

Generation alignment control

First we show the fully centralized tertiary-level control results of aligning the generations of the generating units participating in secondary-level control. The performance objective is chosen as the alignment of all generation ratios, defined as the ratio of reactive generation to the maximum capacity of each generator. The idea is that the generation limit is the lowest possible when the generators are kept aligned with their generations, because all generators reach their limits at the same time. The maximum generation capacities and actual generations of the 9 generators in regions III and IV, before and after tertiary-level control, are given in Table 6.3.

Before tertiary-level control, the reactive power generation outputs are quite uneven, with one generator named LOIRET 3 absorbing reactive power of 83 MVar. Let us plot the generation ratios for the 9 generators in region III and region IV before and after tertiary-level control (see Figure 6.7).

It is seen that the tertiary-level control scheme eliminates the situation of a generator's absorbing reactive power as an actual load to the network. The generator LOIRET 3 is generating reactive power of 94 MVar after the control. The generation ratios are uniformly made more even towards the complete alignment line. It is also noted that the units do not reach full alignment. The reason is that some units are already operating at their limits, and thus no further adjustments on their terminal voltages can be done. However, all units go toward the alignment line after tertiary-level control, even when some units are operating at their limits.

Tie-line flow control

Tertiary-level control can reset the tie-line flows by adjusting the pilot load voltages in each region. This can be done by imposing an additional equality constraint

$$\mathcal{F}[K] = \mathcal{F}^{set} \tag{6.159}$$

where $\mathcal{F}^{set}$ is the desired scheduling value. Using (6.143), one has

$$\mathcal{N}_G \mathbf{V}_G[K] + \mathcal{N}_c \mathbf{V}_c[K] = \mathcal{F}^{set} \tag{6.160}$$

This is already in the form of (6.119) or (6.142) and can be readily incorporated into (6.119) or (6.142). The optimal solutions have thus exactly the same form as the ones given before.

We study the scenario in which the reactive load at bus BOUDES61 in region III increases by 160 MVAR at time $t = 500$ seconds. The goal of tertiary-level control is to increase the reactive tie-line flow from CHAFFS71 in region IV to COULAS71 in region III by 20 MVar to account for the load increase in region III. Fig. 6.8 shows the reactive tie-line flow on this line. The control is activated at time $t = 800$ seconds. Clearly we see that the flow is successfully controlled to the desired new steady-state value.

6.7.7 IAVC

In the secondary-level voltage control practiced by EDF, the control gain parameter is chosen such that the network will settle to a steady-state in three minutes. However, this is not truly guaranteed if there are interactions among neighboring regions that are not taken into account. The IAVC adjusts the secondary-level regulation of the system, taking the change in the tie-line flow into consideration.

However, the initial conditions of the French network modeled by the software developed by the EDF (CODYSIL) do not provide a sufficiently large change over the three minute time in the tie-line flow to show the effect of IAVC. Thus, tertiary-level voltage control is employed at $t = 200$ sec to produce this change. Because of interactions between Region III and Region IV, the system does not settle to steady-state in three minutes ($t = 380$ sec), as one can see in the reactive tie-line flows in Figure 6.9a. When IAVC is applied the dynamic interactions between the two regions are settled in three minutes (Figure 6.9b), and the system has once again more control reserve in the decentralized regions.

IAVC is also effective when reactive exchange support is created by a large load change. A 1667 MW and 1667 MVar load drop is implemented in a simulation and the difference between the case with and without IAVC is obvious (Figure 6.10).

6.7.8 Control at the tertiary level

Base case and TVC case

A simulation was carried out without tertiary voltage control. As can be seen in Figure 6.11a, pilot node CHAFFS71 does not go to its setting (denoted by a gray line) because generators BUGEYT 2 (Figure 6.11b) and LOIRET 3 (Figure 6.11c) reach their terminal voltage limits. When a tertiary control step is applied first at $t = 200$ sec, these generators are taken away from their constraints (Figures 6.12b, 6.12c) and CHAFFS71 (Figure 6.12a) reaches its new setting. In addition, generator LOIRET 3 is now producing instead of absorbing reactive power (Figures 6.11d, 6.12d).

It is important to note that typically two or three tertiary control steps are necessary to bring the network to a steady-state that gives a minimum cost in the performance criterion. Typically a sequence of tertiary control actions enables the pilot voltages to converge to a system optimum suggested by a particular set of nominal values, chosen such that the generator voltages will not exceed their bounds.

In [46] one could find additional simulations showing the comparison of the TVC effects for lower and higher nominal generator voltages. While the generators are taken away from their constraints when these are lower, the pilot-node voltages are generally also lower compared to the case when the nominal generator voltages are higher. This has the implication of preparing the system for different load change contingencies (e.g., high voltage desirable in preparation for load increase).

Furthermore, if the low weight associated with a particular variable in the performance criterion does not provide acceptable regulation at a particular load or generator to recede from its constraint, adaptive tuning of the individual weight in P, Q, or R corresponding to that particular node is necessary to bring the node away from its constraint [46].

6.8 Summary

This chapter presents a structurally based modeling approach for reactive power and voltage dynamics of an interconnected power system. The dynamics of the system are formulated by combining the local dynamics of individual generators and the network couplings. It is shown that the decoupled reactive power/voltage dynamics of power systems are not structurally singular. Quasi-static dynamic models on slower time scales are derived. The structural models developed here are used for systemwide voltage control on slower time scales.

Next, a hierarchical structure of voltage control for large-scale power systems is presented. Dynamics of an interconnected system are formulated by combining the local dynamics of individual generators and the network couplings. Under the assumption of stable primary-level voltage control design, the discrete-time voltage dynamics on slower time scales are derived. Improved secondary-level control is introduced and compared with conventional control. Simulations show a significant improvement over conventional control. The concept of tertiary-level coordination is introduced. Simulations are done for both a small power system and the large-scale French network to illustrate the proposed tertiary-level control schemes.

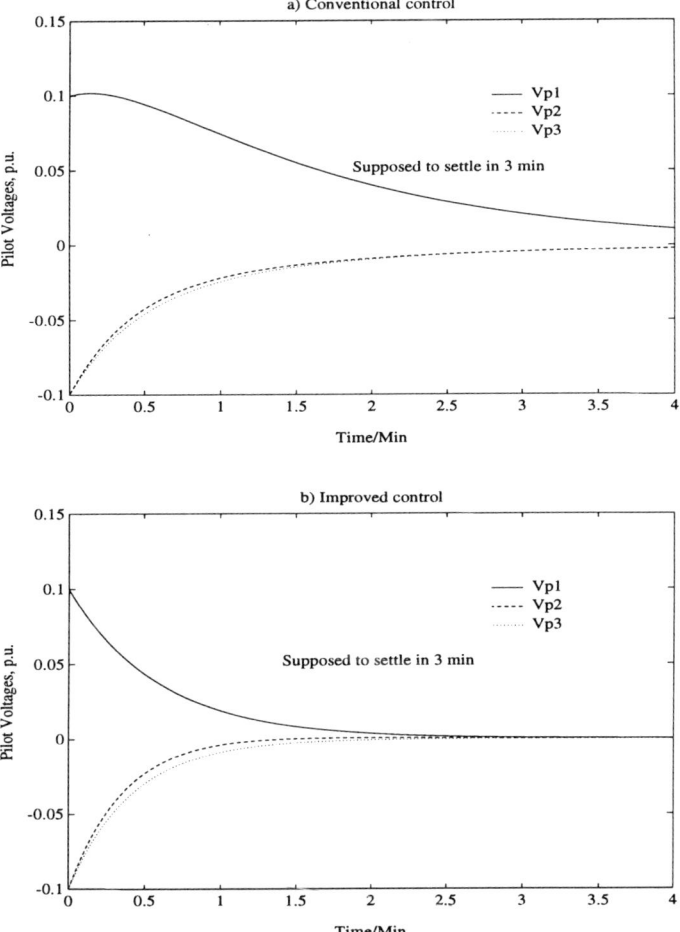

Figure 6.4: Pilot load voltages: (a) conventional, (b) improved

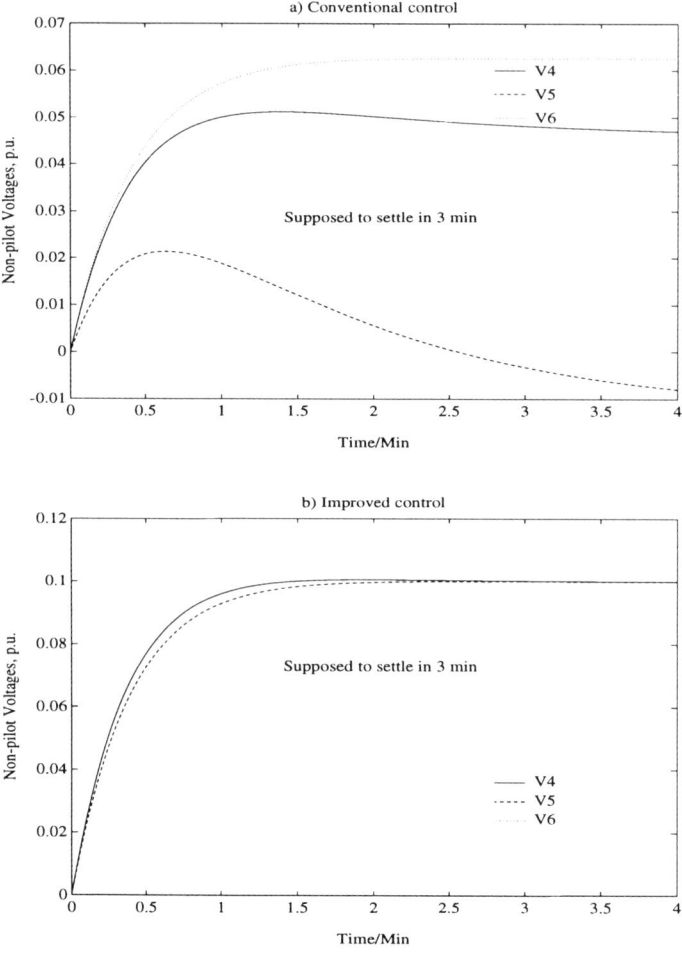

Figure 6.5: Nonpilot load voltages: (a) conventional, (b) improved

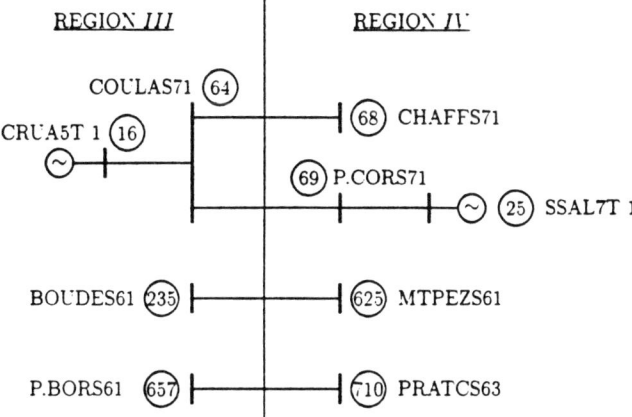

Figure 6.6: Tie-line interconnections on the French network

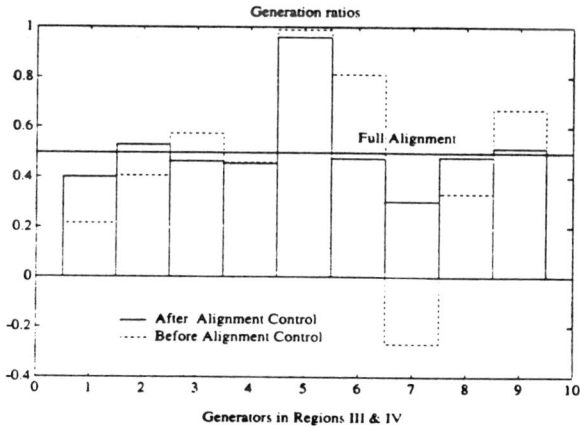

Figure 6.7: Generation ratio alignment

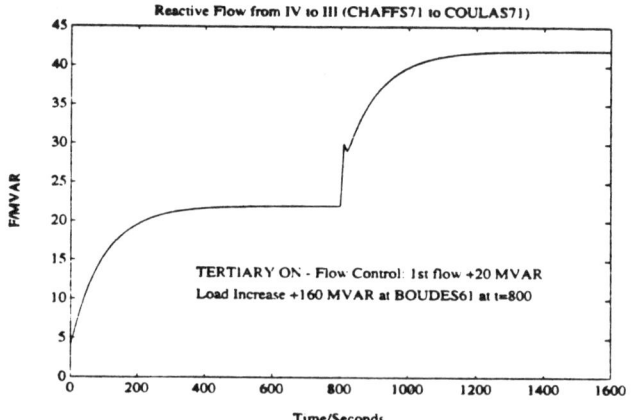

Figure 6.8: Tie-line flow control

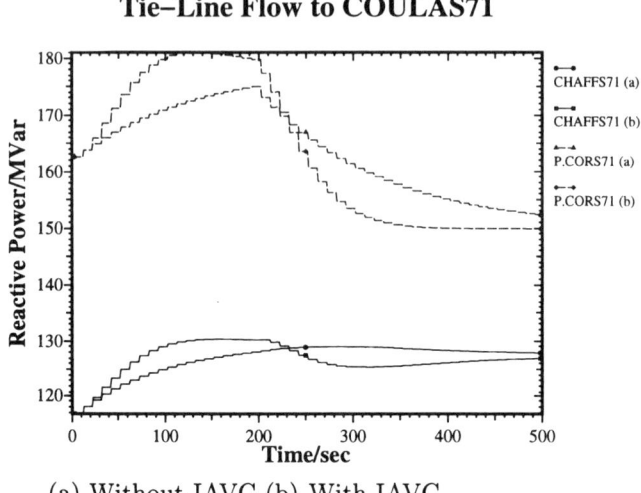

(a) Without IAVC (b) With IAVC

Figure 6.9: The effect of IAVC

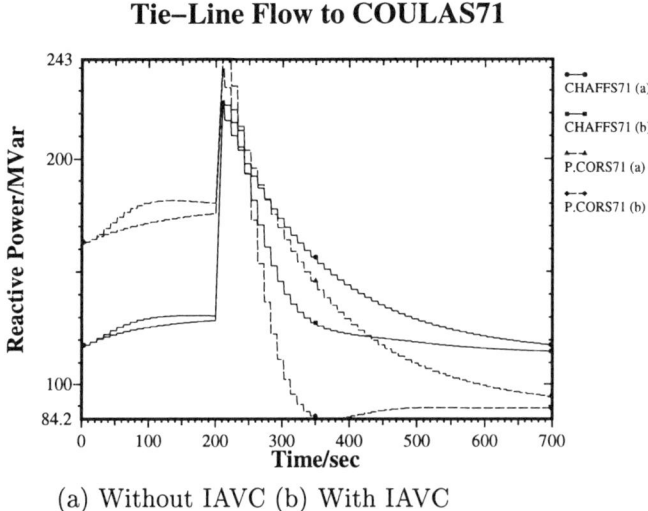

(a) Without IAVC (b) With IAVC

Figure 6.10: IAVC at 1667 MW and 1667 MVAR load drop

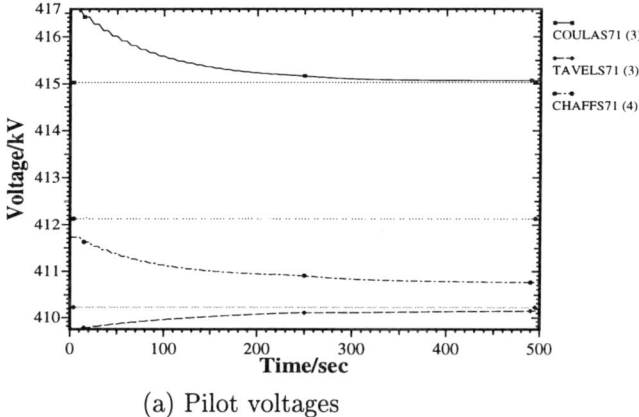

(a) Pilot voltages

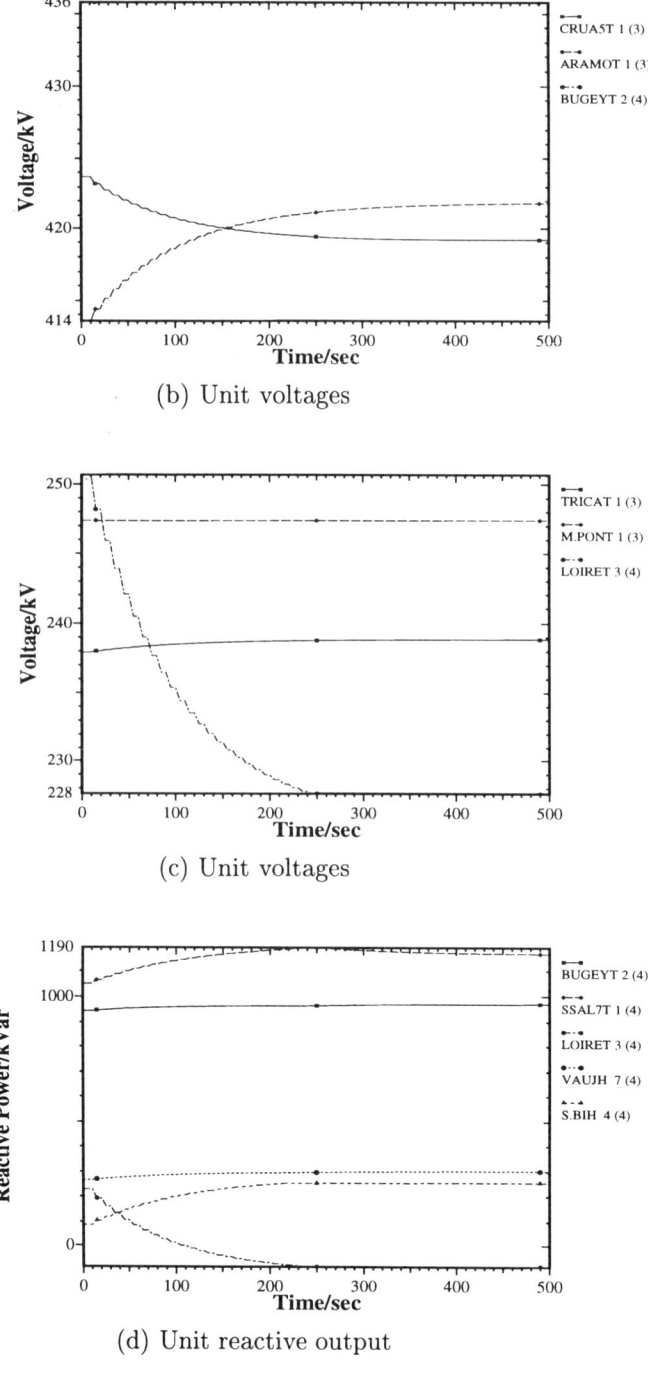

(b) Unit voltages

(c) Unit voltages

(d) Unit reactive output

Figure 6.11: Plain case

Chapter 6

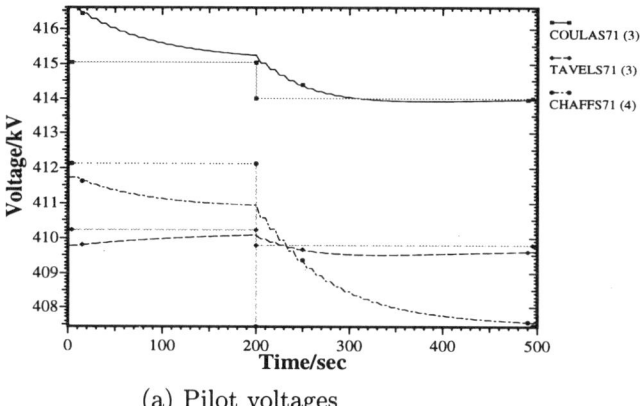

(a) Pilot voltages

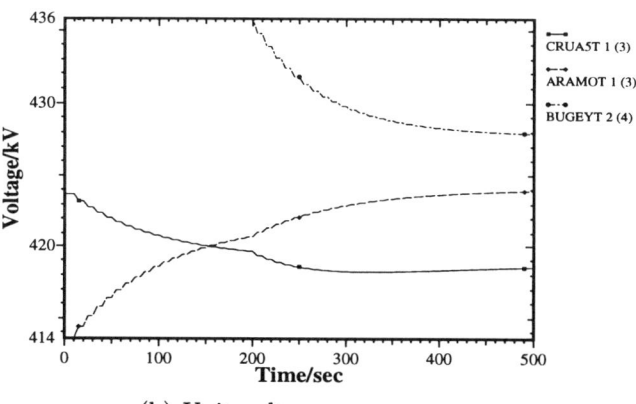

(b) Unit voltages

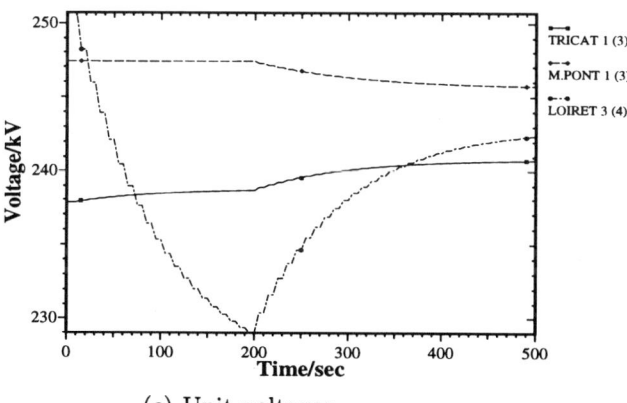

(c) Unit voltages

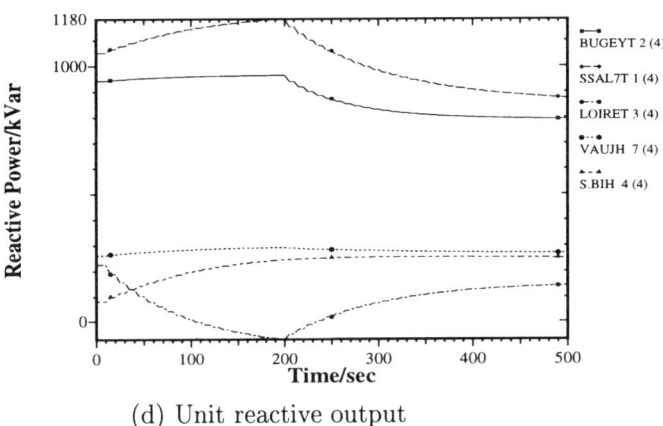

(d) Unit reactive output

Figure 6.12: TVC case

CHAPTER 7
THE VALUE OF GENERATION-BASED REGULATION: COMPETITION VERSUS COORDINATION

In attempting to define a framework for the complex topic of this chapter, we start by recalling the objectives of optimal performance under competitive energy management, described in Chapter 3.

A look at the present electric power industry indicates that much engineering development has gone into efforts to develop control service tools that make possible the use of systemwide resources at the least possible total energy cost while simultaneously meeting technical constraints. Highly technical tools such as economic dispatch (ED), automatic generation control (AGC), and more recently, optimal power flow (OPF) methods are used in many control centers as standard means of facilitating systemwide optimal performance while observing operating constraints. For reliability, only those economically attractive solutions are implemented that are not likely to endanger system integrity when any single unexpected contingency takes place on the system. Each utility controls its activities by observing its basic obligation to serve load as the operating conditions change, including single contingency occurrences. Recall from Chapter 3 that for nominal load there exists an ideally achievable systemwide performance for any given transmission system and its operating constraints. This optimum is referred to as the ideally achievable cost of a given system at its nominal load.

As described in Chapter 3, the objectives of energy management under competition, on the other hand, are for individual subsystems to minimize their own costs without explicitly considering the performance of the interconnected system. The transmission network is assumed to remain capable of accommodating the most attractive economic transactions (i.e., the ones leading to minimum cost for individual subsystems) in spite of the fact that the pattern of flows could differ sharply from the intended design conditions for nominal load. Utilities are also expected to provide systemwide frequency and voltage regulation at a cost that does not reflect system conditions [21, 111]. There is a very strong risk that parties promoting competition will undervalue the role or complexity of the transmission and control services needed to keep the system together despite optimization actions attempted by different competitive parties.

This sharp change of viewpoint, from coordination to competition, leads to two important issues:

1. The achievable optimum (relative to the ideal) strongly depends on the type of industry structure.

2. The right performance-based value of transmission and control services must be determined.

Both issues are supported by rigorous mathematical derivations here. Before carrying out these derivations we formally state what is meant by the term *industry structure* in this text.

- **Definition 7.1:** An (arbitrary) industry structure is defined by its nested hierarchy, as introduced in Chapter 2, and its performance criteria at various levels of this hierarchy, as described in Chapter 3.

The basic setup for analyzing the role of systems control under competition assumes that while competitive market participants (CMPs) are free to make decisions at their own levels, they will be obliged to pay charges for the systems control services, or ancillary services, needed to accommodate their presence on the system. The claim is made that the overall performance of the interconnected system, as measured in terms of criteria defined in Chapter 3, strongly depends on the coordination of controls at various levels of the nested hierarchy and on the price-charging mechanisms that allocate the values of these controls to specific CMPs.

We here suggest a plausible set of general performance criteria for attractive pricing in a competitive environment:

Any optimal price-charging mechanism for systems control services [117] is one that can

1. Encourage technical and economic efficiency at the interconnected system level.

2. Ensure equal treatment for all comparably situated parties.

3. Cover the full capital and operating costs of the party providing transmission and control services.[1]

4. Avoid excessive rules and regulations.

5. Remain relatively stable (nonvolatile) over time.

Criteria 1 and 2 are conventional economic policy standards [52, 56]. What is of interest in the context of the control methodology described in this text is that the ability to satisfy these standards will depend on the management structure of the power industry. A certain amount of coordination appears to be essential for operating efficiency, and centralized accounting for decentralized (self-scheduled) transactions is essential for efficient and equitable cost allocation. In particular, loop flows cannot be readily addressed without usage-based, control-based accounting such as we propose.

Criterion 3, full cost recovery, is of course a condition for long-term economic viability of the transmission providers, and it must be a constraint on prices designed for meeting the other objectives.

Criteria 4 and 5 are very relevant in relation to the approach developed in this text. Any valuation must be done using a minimal yet sufficient information structure, and tools are required that are based on optimal controls over long periods of time. The problem of stability of values (and their associated price components) cannot be addressed by means of static tools such as OPF. Most of the values of control services and the supporting control actions are cumulative in nature over time horizons not typically captured in transient, stability-type, dynamic models of electric power systems. A sufficiently simple control-driven model is needed that captures outputs over long periods of time in response to slowly varying loads or open-access-related discrete events on the system. Such candidate models are introduced in Chapter 4 and illustrated in Chapters 5 and 6 in the context of real and reactive power generation-based systems control.

With these models available one can pose the problem of optimal technical and economic performance as the problem of real-time feedback design in response to input changes caused by various CMPs. The economic feedback should reflect the value of systems control support to the CMPs and should be designed accordingly. The premise is that when these signals are provided in real time, the CMPs can adjust their power quantities traded. If the economic signals are provided for the expected market, a near-optimal price-charging mechanism at a systemwide level is attainable. Conditions under which this general claim can be proven to guarantee systemwide optimal performance can be found in [39, 40]. An illustration of the use of this principle for AGC can be found in [114]. The degree of uncertainty of the evolving market will be a determining factor in how closely the systemwide optimum is followed over different time frames [116].

---
[1] In the case of competitive generation-based systems control, this is no longer required.

We suggest that a certain degree of coordination of systems control services and the accounting for these services is indispensable. The main reason for this claim is strictly technical. Based on the systems control principles described in Chapters 4, 5, and 6, which rely on a hierarchical approach to meeting performance objectives at particular subsystem levels, it appears critical to preserve this general mode of operations. Abandoning these technical principles, and expecting the system to remain intact in response to decentralized economic signals, is highly unrealistic, given our existing technical knowledge. Strictly speaking, no technical solutions exist that are capable of meeting systemwide frequency and voltage quality in a decentralized manner according to the performance objectives described in Chapter 3.

In what follows we state two main points with regard to decentralization: (1) the system may become unstable over mid- and long-term horizons without certain minimal frequency regulation [147], and (2) the decentralized optimization of systems control is inherently suboptimal relative to systemwide optimization. A third, nontechnical, point is that the allocation of charges for systems control resources critical to meeting systemwide performance cannot be done in a fully equitable manner without coordinated accounting. This problem is often illustrated in industry as the loop-flow problem, concerned with accounting for systems control services that accommodate wheeling transactions across large geographical areas [113].

7.1 Relevant optimality questions for determining the value of control services

Using the proposed information structure, which has been shown to be a straightforward generalization of the present structure for automated frequency regulation of a multi-utility interconnected system, we consider the role of control services in providing optimal performance.

Note that decentralized optimization at a secondary level does not necessarily result in systemwide optimal performance. The value of control services is therefore typically different to the interconnected system than to specific subsystems. It is proven rigorously in Section 7.4, in the context of the relevant information structure, that only when the systemwide performance criterion is a direct sum of the performance criteria of individual subsystems, which are in turn functions of control variables, does decentralized optimization as practiced in present power systems lead to the ideally achievable optimum.

An a priori modeling assumption of the static optimization tools used in complex electric power systems (including OPF) is that the real power output from generators $P_G[k]$ is a directly controlled quantity. Since this is not the case, a more careful interpretation of control contributions is needed. Since real power generation is a state variable in a model essential for frequency regulation (Chapter 5), instead of a control variable, optimizing a performance criterion specified in terms of real power generation does not lead to the achievement of a systemwide optimum. Also, one should be particularly careful with the asymmetry caused by inherently singular real power dynamics and the additional burden on controls to prevent the systemwide frequency from floating. If it is important to individual subsystems to have good frequency quality, they must compensate the transmission and control services for this function as well.

One can deduce from the general results in Section 7.4 that under certain conditions the market can become unstable. Using models introduced in this text, we give a general condition reflecting the interdependence of electrical properties of the transmission system and the rules of competition, such as pricing for systems control services, that are sufficient for the market to remain stable. This analysis cannot be done by means of static optimization tools such as OPF. Thus, a competitive market that trades power on the basis of short-run marginal cost (SRMC) spot price estimates using OPF may not be dynamically efficient.

Section 7.5 provides in the context of the relevant information structure a quantification of deviations from the achievable ideal optimum for the industry structure adopted as well as the rules of competition.

The rigorous mathematical facts of Sections 7.4 and 7.5 lead us to conclude that the valuation of transmission management activities by individual subsystems in a competitive market should be strongly related to the technical processes on the interconnected system that are facilitating the network support. We suggest that starting from the already known principles in operating interconnected systems, and unbundling the present information used for multi-utility energy management, much can be achieved toward this goal. Some candidate definitions, essential for linking system operation to market value, are described next. They are applicable to any electric power industry structure, which makes them highly flexible. It is indicated how they can be used for on-line accounting of operating costs and for developing better tools for investment recovery. Investment recovery cannot be done by means of quantities, such as embedded cost, that are used at present in utility bookkeeping.

7.2 Control-dependent values of subsystems in a competitive environment

Consider an interconnected system consisting of several administratively separate subsystems.[2] Under any management structure, the technical value of

---

[2] As described in Chapter 2, the subsystems are not required to map into particular utilities. They could represent any choice of competitive player, including a non-utility-owned generator (NUG). It is indicated at the end of this section that *any* topology, such

participating in energy exchange with other subsystems to any subsystem i, $i = 1, 2, 3$, at time kT_s is given as

$$TV^i[k] = J^i(x^i[k], u^i[k], F^i[k]) - J^i(x^i[k], u^i[k], F^i[k] = 0), \quad i = 1, 2, 3 \quad (7.1)$$

Here F^i is the effective tie-line flow into area i. This quantity is referred to as the technical value of participant i in the energy exchange within an interconnected system at time kT_s. This technical value is exactly the difference between the value of the performance cost in the context of controls whose optimization is the main objective of each subsystem i in a competitive environment within the interconnected system, and the same quantity when the subsystem is disconnected from the rest of the system.

This definition is very general, in the sense that it captures any operating mode at any instant of time k over the long time horizon relevant for cost accounting. Furthermore, it is not biased, since it allows for any subsystem to do anything agreed upon under the adopted industry regulations.

Obviously, at an arbitrarily chosen instant of time k, some subsystems will have their TV^i positive and some negative. For the operating cost pricing component to adequately reflect the operating cost of control actions over periods of time longer than the rate at which the controls are done, a cumulative effect of the above-defined quantity is relevant.

This leads to our second definition. The operating cost component at time KT_t is

$$TV^i[K] = \sum TV^i[k] \quad (7.2)$$

A variety of implementations for economic purposes are possible that would be based on this value, ranging from spot pricing at each $[k]$ through cost evaluation each day or more. The fundamental concept here is that, although the billing need not be done very frequently, the individual subsystems would be required to monitor this quantity at the rate the controls are done. This would be simple to do, since only values of local (subsystem-level) states and controls as well as locally measurable effective tie-line flows are required.[3] It is pragmatic to account for these values over long time horizons. This leads to a useful definition for pricing purposes.

The $TV^i[K]$ as defined here are entirely (performance) usage-based and can account for any hard operating and control constraints as long as the performance criterion $J^i[k]$ includes all technical specifications at a subsystem level. Under unusual circumstances it may happen that subsystem i is not

---

as a NUG within one utility interconnected to others, lends itself naturally to the concepts proposed in this section.

[3] One consequence of the proposed formulation is the notion of effective tie-line flow instead of net tie-line flow. This is important, because it accounts for electrical distances at a subsystem level.

Chapter 7   175

capable of functioning at all without significant help from neighbors. A simple example of this is a generation outage in this subsystem. The same definitions apply to this situation.

7.3 Systems control structure-related issues

While the preceding proposed performance-based definitions universally hold, since they are independent of the type of management or industry structure, their use for allocating the computed value at a subsystem level i among all subsystems is a much harder question. The answer depends on the market structure adopted. Primarily because the results of decentralized optimization are not the same as the results of centralized optimization, the answers are different.

The cleanest setup would be to have a centralized management responsible for tertiary-level scheduling of tie-line flows among the subsystems to allow for near-ideal performance of the interconnected system in order to meet the first objective of optimal pricing efficiency. This mode is particularly attractive, especially given the derivations in Section 7.4 showing that only some very narrow classes of performance criteria allow for this to happen without centralization.[4] This leads to two, qualitatively different, options.

Centralized structure for systems control services in a competitive environment

Based on the proposed measure of systems control value, it is possible to directly monitor performances of individual subsystems by means of a generalized AGC-like information structure and to compute their cumulative changes in a decentralized manner using the general modeling approach in this text.

Centralization becomes essential, however, for systemwide frequency regulation [5] and for computing adequately the contributions of individual subsystems to system performance changes over time for open access and the presence of NUGs. In other words, equal treatment of similar parties and full cost recovery can be guaranteed only under centralized energy management.[6] It is arguable, based on the derivations in this text, that a straightforward formula relating all individual line flows to the generation-based controls in

---

[4] For minimal regulation, see [147].

[5] As described in Chapter 6, a similar information structure could be adopted for voltage regulation [74].

[6] Therefore, in our recent correspondence to the Federal Energy Regulatory Commission (FERC) [76], a notion of centralized energy management was introduced. Note that a variety of ex post mechanisms for correcting non-cost-recovering prices could be devised, but these would not be optimal by virtue of violating criterion 1, i.e., overall technical efficiency.

particular subsystems can be developed that is directly useful for this purpose.

The general structure of such a formula is given in Section 7.4, Eqs. (44) and (51). Further details of this relation can be found in [17, 75]. This formula represents an aggregate equality constraint at the tertiary level. It identifies, in a very concise way, by means of a reduced information structure, the relative contribution of any subsystem j to the change in performance criterion of any subsystem i. This is all that is needed for resolving the parallel path problem as posed at present.

It is important to appreciate the role of discrete models defined here for such evaluation; the explicit relation between flows and controls is a direct result of this modeling. The relative contribution of individual subsystems to values like $TV^i[K]$ will have to be computed in a centralized (tertiary-level) manner at each discrete instant k when controls at a subsystem level act. While only a modest information set is required, there is sufficient detail for cost recovery pricing that is allocated down to individual transactions as frequently as idealized real-time spot pricing. However, such allocation costs can be collected over any convenient billing horizon; spot pricing is not required.

Decentralized structure for systems control services

This type of implementation, although much promoted, must be handled with great caution in order to avoid potential chaos in transmission and control services of an electrically interconnected system.

Recall from Chapter 5 that at present we do not have a technical solution for fully decentralized frequency regulation. AGC does not guarantee that the absolute frequency is maintained at its nominal value without additional time correction at one generator designated to perform this function at the interconnected system level [4]. This generator effectively maintains the system reference angle at zero, no matter what else happens on the system. This function, however, implies a certain cost of generation at the interconnected system level, which is not easily allocated to lower subsystem levels.

This problem is not insoluble, but it will require more technical research. One possible solution is to have all generation-based controls receive the signal from the global positioning system (GPS) satellite and regulate their frequencies relative to this signal. This would potentially eliminate the inherent asymmetry that exists in the present mechanism of maintaining the system reference angle at zero by a generator that happens to be geographically located in one of the subsystems. The cost of additional communications and control adjustments needed to implement this solution should be carefully evaluated, if this becomes an option. For the time being we observe that there is an additional burden beyond the AGC function necessary to guarantee that the absolute frequency to remain close to its nominal value.

Without this technical activity the interconnected system generally exhibits a drift in absolute frequency that may become detrimental to system integrity; without this function the under-frequency protection would disconnect generators as the absolute frequency exceeds the prespecified safe limits, and this would lead to a potential sequential disintegration of the interconnected system in the form of partial or full blackouts. It should be clear that without a control-based solution it is impossible to value the contributions of individual subsystems to the technical value under competition of the ith subsystem.[7]

Even if this technical obstacle is successfully resolved, the question of potential suboptimal use of systemwide energy resources (utility as well as non-utility-owned) remains a serious one. One must study this issue in the context of optimal pricing. The results in Section 7.5 provide starting formulae for quantifying deviations in performance under different management structures. These formulae should be studied in the context of representative utility data. These technical problems must be resolved before decentralized competitive energy management can be adopted with confidence.

7.4 Long-term stability of decentralized systems control services

The formulation and proofs in this section are needed for differentiating the way system optimization in a deregulated utility environment is formulated at present from the most likely ways it will be formulated in the future. These issues are typically not well thought out, and their implications not fully appreciated. As a result, much confusion arises in interpreting how future energy management should be done.

Consider any administratively separate region i within an interconnected system. Let the control variables be u^i (real power generation settings or voltage settings), and the state variables be x^i (frequencies and voltages of loads). The task of controllers at a subsystem level is to determine u^i in order to minimize a cost function:

$$J^i = J^i(x^i, u^i) \tag{7.3}$$

where $i = I, II, \ldots, R$, assuming there are R regions in the system.

Using our structure-based modeling approach, the states x^i and controls u^i at each subsystem level are shown to be related as

$$x^i = A^i u^i + B^i F^i + L_0^i \tag{7.4}$$

where A^i, B^i, L_0^i are constants, and F^i represents the tie-line flows into region i.

---

[7] A recent illustration of this situation was the process of rolling blackouts in the Northeast during the cold winter of 1994. This scenario must be analyzed.

The result of this decentralized optimization process can be written as

$$u^i = f^i(F^i) \tag{7.5}$$

On the interconnected system level, with the notation

$$\mathbf{x} \triangleq \begin{bmatrix} x^I \\ \vdots \\ x^R \end{bmatrix} \quad \text{and} \quad \mathbf{u} \triangleq \begin{bmatrix} u^I \\ \vdots \\ u^R \end{bmatrix} \tag{7.6}$$

$$\mathbf{F} \triangleq \begin{bmatrix} F^I \\ \vdots \\ F^R \end{bmatrix} \quad \text{and} \quad \mathbf{f} \triangleq \begin{bmatrix} f^I \\ \vdots \\ f^R \end{bmatrix} \tag{7.7}$$

(7.5) rewritten in a vector form for all regions takes on the form

$$\mathbf{u} = \mathbf{f}(\mathbf{F}) \tag{7.8}$$

The updating of the settings $\mathbf{u}$ is done only at discrete times kT, where T is the interval for updating the new $\mathbf{u}$. Let us denote any quantity at discrete times kT as $\mathbf{u}[k] \triangleq \mathbf{u}[kT]$. Eq. (7.5) can be rewritten as

$$\mathbf{u}[k] = \mathbf{f}(\mathbf{F}[k]) \tag{7.9}$$

Data at time kT are used to compute controls for the next time $(k+1)T$. This leads to a dynamic process:

$$\mathbf{u}[k+1] = \mathbf{f}(\mathbf{F}[k]) \tag{7.10}$$

The equilibrium points and stability over the long-term horizons of this dynamic process can be studied by utilizing the relation between the flows and the states, obtained through an aggregate model,

$$\mathbf{F} = \mathbf{N}\mathbf{x} \tag{7.11}$$

where

$$\mathbf{N} = \begin{bmatrix} N^{I,I} & \cdots & N^{I,R} \\ \cdots & \cdots & \cdots \\ N^{R,I} & \cdots & N^{R,R} \end{bmatrix} \tag{7.12}$$

is the coupling matrix with nonzero off-block diagonal elements.

This formula is the basis for solving the parallel path problem, since it defines the relative contribution of states participating in systems control in particular areas.

Chapter 7

To write the dynamic process in terms of flow vector $\mathbf{F}$ only, let us denote further

$$\mathbf{A} \triangleq \text{Block diag}\{A^I, \ldots, A^R\} \tag{7.13}$$

$$\mathbf{B} \triangleq \text{Block diag}\{B^I, \ldots, B^R\} \tag{7.14}$$

With this notation, we can write the constraints as

$$\mathbf{x} = \mathbf{Au} + \mathbf{BF} + \mathbf{L}_0 \tag{7.15}$$

Together with (7.11), this leads to

$$\mathbf{F} = \mathbf{N}(\mathbf{Au} + \mathbf{BF} + \mathbf{L}_0) \tag{7.16}$$

or

$$(\mathbf{I} - \mathbf{NB})\mathbf{F} = \mathbf{NAu} + \mathbf{NL}_0 \tag{7.17}$$

or

$$\mathbf{F} = (\mathbf{I} - \mathbf{NB})^{-1}\mathbf{NAu} + (\mathbf{I} - \mathbf{NB})^{-1}\mathbf{NL}_0 \tag{7.18}$$

Then we have

$$\mathbf{F}[k+1] = (\mathbf{I} - \mathbf{NB})^{-1}\mathbf{NAu}[k+1] + (\mathbf{I} - \mathbf{NB})^{-1}\mathbf{NL}_0 \tag{7.19}$$

Using (7.10), one obtains the explicit dynamic process in terms of flow only,

$$\mathbf{F}[k+1] = (\mathbf{I} - \mathbf{NB})^{-1}\mathbf{NAf}(\mathbf{F}[k]) + (\mathbf{I} - \mathbf{NB})^{-1}\mathbf{NL}_0 \tag{7.20}$$

The equilibrium point is given by

$$\mathbf{F}_\infty = (\mathbf{I} - \mathbf{NB})^{-1}\mathbf{NAf}(\mathbf{F}_\infty) + (\mathbf{I} - \mathbf{NB})^{-1}\mathbf{NL}_0 \tag{7.21}$$

The stability of the equilibrium point is determined by the eigenvalues of the matrix

$$(\mathbf{I} - \mathbf{NB})^{-1}\mathbf{NA} \left.\frac{\partial \mathbf{f}}{\partial \mathbf{F}}\right|_{\mathbf{F}_\infty} \tag{7.22}$$

If all the eigenvalues of the matrix in (7.22) are inside the unit circle, then the dynamic process is stable. Otherwise, it is unstable.

This matrix is seen as the measure of the interconnection strength among subsystems.

7.5 Achievable optimality as a function of the level of control coordination

In this section, a mathematical formulation needed to quantify systemwide deviations from the ideally achievable optimum is proposed. The necessary optimality conditions are strongly dependent on the type of information exchange available. Case A provides an interpretation of a centralized optimum in terms of flow sensitivities explicitly. Case B is concerned with different implementations of decentralized structures for systems control services.

Case A—centralized systems control

Assume that the centralized cost function is the sum of all regional cost functions, i.e.,

$$J = \sum_{i=I}^{R} J^i \tag{7.23}$$

and that all information needed for centralized optimization is available for all regions. One has

$$\frac{dJ}{du^j} = \frac{\partial J^j}{\partial u^j} + \sum_{i=I}^{R} \frac{\partial J^i}{\partial u^j} \tag{7.24}$$

$$= \frac{\partial J^j}{\partial u^j} + \sum_{i=I}^{R} \sum_{k=I}^{R} \left[\frac{\partial x^k}{\partial u^j}\right]^T \frac{\partial J^i}{\partial x^k} \tag{7.25}$$

$$= \frac{\partial J^j}{\partial u^j} + \sum_{i=I}^{R} \left[\frac{\partial x^i}{\partial u^j}\right]^T \frac{\partial J^i}{\partial x^i} \tag{7.26}$$

Note the presence of the cross-terms $\partial x^i / \partial u^j$, which can be obtained from

$$x^i = A^i u^i + B^i F^i + L_0^i \tag{7.27}$$

$$\frac{\partial x^i}{\partial u^j} = \begin{cases} A^i + B^i \frac{\partial F^i}{\partial u^i} & j = i \\ B^i \frac{\partial F^i}{\partial u^j} & j \neq i \end{cases} \tag{7.28}$$

The term $\partial F^i / \partial u^j$ can be calculated from (7.17) as

$$\frac{\partial \mathbf{F}}{\partial \mathbf{u}} = (\mathbf{I} - \mathbf{NB})^{-1} \mathbf{NA} \tag{7.29}$$

Chapter 7

Case B—decentralized systems control

In the decentralized case, each region optimizes its own cost function J^i. We can show that

$$\frac{dJ^i}{du^i} = \frac{\partial J^i}{\partial u^i} + \left[\frac{\partial x^i}{\partial u^i}\right]^T \frac{\partial J^i}{\partial x^i} = 0 \tag{7.30}$$

Using (7.28), one obtains

$$\frac{\partial J^i}{\partial u^i} + \left[A^i + B^i \frac{\partial F^i}{\partial u^i}\right]^T \frac{\partial J^i}{\partial x^i} = 0 \tag{7.31}$$

Comparing with the centralized method, we see that all

$$\frac{\partial x^i}{\partial u^j} \quad j \neq i \tag{7.32}$$

are neglected in the regional optimizations.

Different ways of treating $\partial F^i / \partial u^i$ yield different results. We propose two strategies.

Constant-flow strategy

In this strategy, flows are viewed as a measured parameter; thus, all derivatives of the flow with respect to controls are neglected, i.e.,

$$\frac{\partial F^i}{\partial u^j} = 0 \quad \forall i, j \tag{7.33}$$

Eq. (7.31) gives, in this case,

$$\frac{\partial J^i}{\partial u^i} + [A^i]^T \frac{\partial J^i}{\partial x^i} = 0 \quad \forall i \tag{7.34}$$

This leads to the solution

$$u^i = f^i(F^i) \tag{7.35}$$

Nash strategy

In this game strategy, flows are viewed as a function of the controls within the region. As a result, all cross-derivatives of the flow with respect to controls in other regions are neglected—equivalent to the so-called reaction curves.

Neglecting $\partial x^i/\partial u^j$, $j \neq i$, we have

$$\frac{\partial F^i}{\partial u^i} = N^{i,i} \frac{\partial x^i}{\partial u^i} \qquad (7.36)$$

Using (7.28), one obtains

$$\frac{\partial F^i}{\partial u^i} = N^{i,i}(A^i + B^i \frac{\partial F^i}{\partial u^i}) \qquad (7.37)$$

or

$$(I - N^{i,i}B^i)\frac{\partial F^i}{\partial u^i} = N^{i,i}A^i \qquad (7.38)$$

or

$$\frac{\partial F^i}{\partial u^i} = (I - N^{i,i}B^i)^{-1} N^{i,i} A^i \qquad (7.39)$$

From this, we obtain the regional Nash optimization scheme from (7.30) as

$$\frac{\partial J^i}{\partial u^i} + \left[A^i + B^i(I - N^{i,i}B^i)^{-1} N^{i,i} A^i\right]^T \frac{\partial J^i}{\partial x^i} = 0 \qquad (7.40)$$

The solution can be written as

$$u^i = f^i(F^i) \qquad (7.41)$$

Note that the two solutions in (7.31) and (7.41) are different in general. Their relation is easily seen as that of letting $N^{i,i} = 0$ in the Nash solution yield the constant-flow solution.

The mathematical derivations in this section are essential for understanding the limitations of decentralized systems control. Only in the very particular case when individual performance criteria are functions of controls will the decentralized and centralized optimizations in a multi-utility environment lead to the same optimum value. This means that a purely decentralized setup will generally result in a deviation of the interconnected system from its ideal optimal performance. *This is a very strong argument for retaining coordinated systems control.*

7.6 Limitations of existing systems control in a competitive environment

In order to meet specified performance criteria, the system operator at present relies primarily on static network and generation modeling tools, such as load flow studies, economic dispatch simulations, or OPF analyses. Such models assume that the state of the system is known with certainty. Of course, the state is not known with certainty, but the assumption is workable given that many of the system functions are under central control and that utilities still have an obligation to serve the needs of their service territories. More specifically, under present conditions:

- Nearly all generating units are scheduled by their respective operating utilities and most are available for coordinated participation in system functions such as load following, frequency regulation, voltage control, and contingency (security) control.

- The forecasted variables (load, unit outage statistics) are reasonably predictable, at least enough so that a system operator can rely on static modeling tools.

In a competitive environment, these assumptions and conditions cannot be relied on. There may be resources that supply power but that may not be able to participate in some systems control functions (particularly those functions related to voltage control and some operating reserves that can only be supplied locally to be effective). Other resources may prefer not to participate. They may choose instead to operate in roughly the same fashion every day (e.g., generating at unity power factor and per-unit voltage of 1.0) and to concentrate on just selling power. Both generation and load resources may jump in and out of the competitive market on an opportunistic, profit-driven basis. Increased uncertainty in the state variables will require more sophisticated look-ahead calculations and dynamic modeling tools. For such reasons, the system operator of the future may have a much tougher job.

Indeed, the system operator of today will need to evolve into an "independent system operator" (ISO), situated at the tertiary level, whose responsibility it is to coordinate events in two markets: the competitive market (CM), comprising individual transactions between competitive market participants (CMPs), and the services market (SM), composed of services market participants (SMPs) providing generation and transmission (ancillary) services to keep the system together and to ensure systemwide performance quality.

7.7 Proposed approach to real-time systems control and its pricing in a competitive market

Based on our analysis here and recalling the types of system inputs, described in Section 3.1, that are likely to require systems control support, we suggest in what follows an approach to real-time systems control and its pricing in a competitive market. This requires

- Efficient creation of systems control services over different time frames to meet specified performance criteria for the anticipated dynamics of CMP transactions.

- Systems control implementations that guarantee systemwide performance.

- Providing meaningful price-charging mechanisms for systems control to the CMPs.

If these are done systematically, a changing industry is likely to achieve success as measured by societal benefits rather than by benefits solely to specific subsystems in the nested hierarchy. Our approach is based on a gradual evolution of present systems control to take into account that system inputs include not only native load demand but also the transactions of profit-driven CMPs.

We have suggested that the provision of ancillary services to keep the system together and the pricing of these services should be coordinated at the tertiary level, taking into account the actions of CMPs. This leads to our proposed iterative integration of these processes.

Ancillary services are needed in part because of transmission system constraints. But even the ISO cannot know precisely where those constraints will be or how severe they may become prior to evaluation of transactions proposed by the CMPs. Thus, some degree of economic decoupling between the ISO activities and the CM is inevitable. However, it is natural to consider an iterative approach to resolving the need for simultaneous information about transactions and about system conditions.

The main reason we suggest real-time implementation of this sort is that it brings both worlds together—the proponents of competition and of centralized pricing.

The integration of operations and pricing is proposed to

- Allow for competitive supply/demand.

- Allow for coordinated management and pricing of generation-based systems control services.

- Allow market participants to switch in time from being CMPs to being SMPs, and vice versa, at will.

Figure 7.1 illustrates this integration in an iterative two-level bidding market.

The arrangement is basically iterative in the sense that the CMPs' interactions are anticipated by the ISO, and that pattern of interactions determines how the SMPs will be used. In turn, the SM costs are charged to and recovered from the CMPs. By sending and receiving relevant signals

Chapter 7

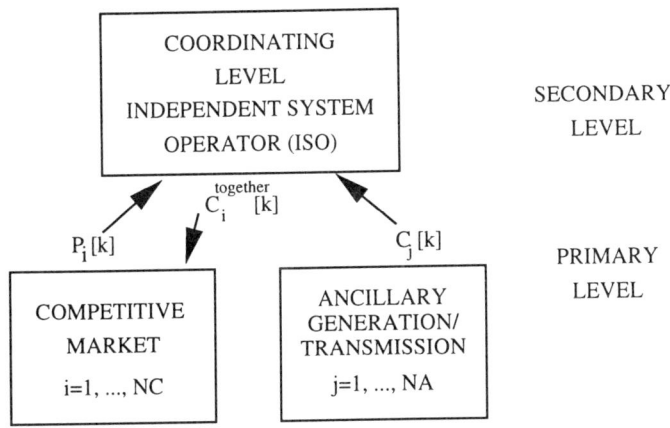

Figure 7.1: Iterative process of accommodating CM from the ISO level

to and from both markets, the ISO ensures that the trade-off between more CM transactions and higher SM charges is brought to an equilibrium. The CM is distributed, with the price terms and conditions of its deals between CMPs remaining proprietary. Only the power quantities and locations of the deals are communicated to the ISO. In contrast, the SM is a coordinated function that conveys a nondiscriminatory price for ancillary services to each CMP, according to (at least) its relative impact on system reliability. This requires new administrative procedures for communication between the ISO and the CMPs/SMPs as well as new computational tools for determining the marginal impact of each transaction on the system.

7.7.1 Basic steps for linking technical and pricing processes

In this and following sections, we use the term *services market (SM)* to mean utility-owned and non-utility-owned generators taking part at the tertiary level in providing systemwide transmission and ancillary services for the benefit of all participants in the system.[8]

In Figure 7.2, a set of asynchronous contracts of various durations is superposed on the time scales associated with secondary- and tertiary-level technical activities.

In order to relate systems control technical processes to contractual processes over the period of time relevant for each individual contract, we propose here that the price for ancillary services should be established in several steps:

1. At the SM (tertiary) level, control actions (3.5) and (3.6) are designed to optimize systemwide performance over the entire time horizon $T_{contract}$

---

[8]This includes both CMPs and the demand not necessarily responsive to price.

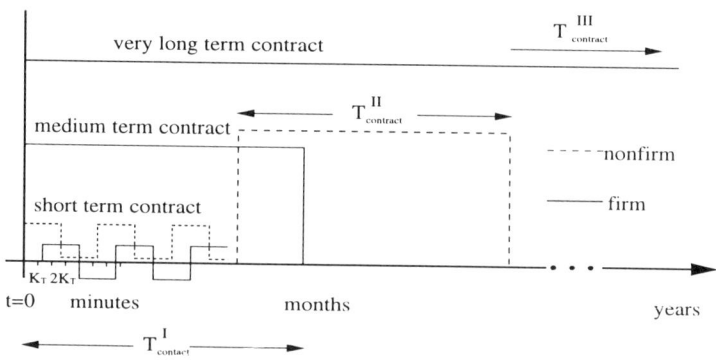

Figure 7.2: Time scales and superposed contracts

according to the best information on expected values for the system condition at the time the contract is established.[9]

2. At the SM level, an estimation process is established for determining the expected economic value of controls (3.5) and (3.6) to the specific parties i_1 and i_2 involved in the contractual agreement over the time $T_{contract}$.

3. The price to these two parties for transmission and ancillary services of the entire systems control services market is determined according to this economic value.

4. The contribution of particular controls members of N_A, to this economic value is determined according to the formulae proposed in the next section.

7.7.2 Operations planning for the anticipated contract at the tertiary level

The tertiary-level control actions (3.6) should be planned for and taken simply by responding to the expected value of new contract-imposed inputs at buses i_1 and i_2 in the same manner as is done at present in response to anticipated load and generation changes in native load and generation.[10] For example,

---

[9]Note that AGC, AVC, and primary controls at each level react in an automated way to the schedules set at the tertiary level.

[10]This example is for real power scheduling. We have developed similar performance criteria and formulae for reactive power scheduling at the tertiary level.

Chapter 7

this calls for designing (3.6) in such a way as to optimize

$$J^{tertiary}_{T_{contract}} = \sum_{K=1}^{\frac{T_{contract}}{T_t}} \underline{P_G}^T[KT_t]R\underline{P_G}[KT_t] + \underline{\omega}^T[KT_t]Q\underline{\omega}[KT_t] \tag{7.42}$$

subject to standard controls

$$P_{Gi}^{min} \leq P_{Gi} \leq P_{Gi}^{max} \tag{7.43}$$

and security

$$P_{ij}^{min} \leq P_{ij} \leq P_{ij}^{max} \tag{7.44}$$

$$\theta_{ij}^{min} \leq \theta_{ij} \leq \theta_{ij}^{max} \tag{7.45}$$

constraints.[11] A similar control sequence is generated according to certain performance criteria relevant for voltage quality.

An efficient technical/economic (cost-based) solution to this problem requires knowledge of the expected values of all system inputs (market-driven and others) over the the contract time $T_{contract}$.

7.7.3 Estimating at the tertiary level the economic value of systems control services to the contract participants i_1 and i_2

For notational purposes, assume bus i_1 to be on the supply side and bus i_2 on the demand side.[12] The typical cost function for generation supply is

$$c_{i1}(P_{Gi1}) = a_{11} + a_{12}P_{Gi1} + a_{13}P_{Gi1}^2 \tag{7.46}$$

Similarly, the typical benefit function on the demand side is

$$b_{i2}(P_{Di2}) = b_{11} + b_{12}P_{Di2} \\ + b_{13}P_{Di2}^2 \tag{7.47}$$

These two functions are referred to as economic value functions relevant for pricing. They will have components contributed by the primary controls at the fast rate t, and components contributed by automatic actions such as AGC at the secondary and tertiary levels at the rates kT_s and KT_t.

For meaningful pricing it is essential to estimate these components at the tertiary level and establish pricing mechanisms accordingly.

---

[11]This is not an exhaustive list of constraints; it is also often specific to a particular system.
[12]For full generalization of these derivations, it is necessary to have a combination of supply and demand at each bus.

7.7.4 Pricing for systems control at the ISO (tertiary) level to accommodate transaction $i_1 - i_2$

The market price for transmission and ancillary services to player i_1 over the contractual period $T_{contract}$ is

$$Price(T_{contract,i1}) = \Xi_{T_{contract,i1}}(-\Delta c_{i1}(P_{Gi1})) \tag{7.48}$$

Similarly, the market price for transmission and ancillary services to player i_2 over the contractual period $T_{contract}$ is

$$Price(T_{contract,i2}) = \Xi_{T_{contract,i2}}\Delta(b_{i2}(P_{Di2})) \tag{7.49}$$

Δ stands for incremental charge during the entire contract duration $T_{contract}$, and it is contributed to by the transmission and ancillary services at all three control levels (primary, secondary, and tertiary). Ξ stands for the expected value. This solution is qualitatively different for any reasonable contract duration $T_{contract}$ from the localized marginal-cost-based solution only at the beginning of contract. It is also qualitatively different from any average-cost, peak-load solution.

This is the economically most efficient solution[13] that one can determine under uncertainties; the longer $T_{contract}$, the more deviations will be seen from optimal system performance. There is potentially a serious risk in committing to a firm fixed price over very long contractual horizons, in terms of inefficient reserve use—a dilemma of guaranteed quantity versus higher price. Thus, algorithms for minimizing this risk should be studied as more is learned about the probabilities of market behavior.

7.7.5 Contribution of individual components to the economic values at buses i_1 and i_2

This is the problem of "dividing the pie" among the members of a nested hierarchy. Consider the following straightforward relation:

$$\Delta c_{i1}(P_{Gi1}) = \frac{\partial c_{i1}}{\partial P_{Gi1}}(\frac{\partial P_{Gi1}}{\partial u_p}\Delta \underline{u}_p + \frac{\partial P_{Gi1}}{\partial u_s}\Delta \underline{u}_s[kT_s] + \frac{\partial P_{Gi1}}{\partial u_s}\Delta \underline{u}_s[KT_t]) \tag{7.50}$$

Here $\Delta \underline{u}_p$ stands for the incremental control actions taken during the entire contract duration by the primary controllers. Similarly, $\Delta \underline{u}_s[kT_s]$ stands for incremental control actions at the secondary level in an attempt to maintain interface flows as scheduled by the pool. Finally, $\Delta \underline{u}_s[KT_t]$ stands for the incremental actions of controls initiated at the tertiary level in anticipation of the contract.

---

[13] In the sense of making the entire system performance maximize social welfare.

Chapter 7

These incremental values are simple algebraic sums of controls at each instant t, $[kT_s]$ and $KT_t]$ respectively.

Similarly,

$$\Delta b_{i2}(P_{Di2}) = \frac{\partial b_{i2}}{\partial P_{Di2}}(\frac{\partial P_{Di2}}{\partial u_p}\Delta u_p + \frac{\partial P_{Di2}}{\partial \underline{u_s}}\Delta \underline{u_s}[kT_s]$$
$$+ \frac{\partial P_{Di2}}{\partial \underline{u_s}}\Delta \underline{u_s}[KT_t] \quad (7.51)$$

The expected value Ξ of these quantities relevant for price setting (7.48) and (7.49) is simply

$$\Xi_{T_{contract,i1}}\Delta c_{i1}(P_{Gi1}) = \frac{\partial c_{i1}}{\partial P_{Gi1}}(\frac{\partial P_{Gi1}}{\partial u_p}\Xi\Delta u_p + \frac{\partial P_{Gi1}}{\partial \underline{u_s}}\Xi\Delta \underline{u_s}[kT_s] +$$
$$\frac{\partial P_{Gi1}}{\partial \underline{u_s}}\Xi\Delta \underline{u_s}[KT_t]) \quad (7.52)$$

Similarly,

$$\Xi_{T_{contract,i2}}\Delta b_{i2}(P_{Di2}) = \frac{\partial b_{i2}}{\partial P_{Di2}}(\frac{\partial P_{Di2}}{\partial u_p}\Xi\Delta u_p + \frac{\partial P_{Di2}}{\partial \underline{u_s}}\Xi\Delta \underline{u_s}[kT_s] +$$
$$\frac{\partial P_{Di2}}{\partial \underline{u_s}}\Xi\Delta \underline{u_s}[KT_t]) \quad (7.53)$$

These values are very difficult for regulators to determine. They are fairly straightforward to determine and monitor at the tertiary level. It is important to recognize next that only u_p, $\underline{u_s}[kT_s]$ and $\underline{u_s}[KT_t]$ for frequency regulation by real power scheduling have significant (fuel-based) operating costs $c_{ij}(\Delta u_p)$, $c_{ij}(\Delta \underline{u_s}[kT_s])$, and $c_{ij}(\Delta \underline{u_s}[KT_t])$, for any generating unit j in region i participating in ancillary services. Similar services for voltage support and transmission service are basically free of operating cost.

Because of this fact, an additional complication arises when working out equitable formulae for profit distribution:

1. Subtract $\sum_i \sum_j (c_{ij}(\Delta u_p) + c_{ij}(\Delta \underline{u_s}[kT_s]+) + c_{ij}(\Delta \underline{u_s}[KT_t])$ from the total profit (7.48) obtained from the market player i_1.

2. Divide the leftover profit from player i_1 according to the sensitivity factors in formula (7.52) so that primary controls u_p make profit proportional to $\underline{p_{i1,p}} = \frac{\partial P_{Gi1}}{\partial u_p}$, secondary-level controls $\underline{u_s}[kT_s]$ make profit proportional to $\underline{p_{i1,s}}[kT_s] = \frac{\partial P_{Gi1}}{\partial \underline{u_s}}[kT_s]$, and tertiary-level ancillary actions $\underline{u_s}[KT_t]$ make profit proportional to $\underline{p_{i1,s}}[KT_t] = \frac{\partial P_{Gi1}}{\partial \underline{u_s}}[KT_t]$.

3. The compensation to transmission services by all lines whose reactances are X_{ij} should be proportional to $p_{i1,X_{kl}} = \frac{\partial P_{Gi1}}{\partial X_{kl}}$ for each transmission line connecting two arbitrary buses k and l, so that for all transmission lines one obtains vector $\underline{p_{i1,X}} = [p_{i1,X_{1-2}} \vdots p_{i1,X_{2-3}} \vdots \cdots]$.

4. This is normalized further by a common denominator α, so that

$$\underline{p_{i1}}^T \underline{1} = 1 \tag{7.54}$$

5. This is a participation formula for distributing the contributions of all ancillary services in an equitable manner.

6. An identical process should be applied for distributing benefits from any other market player (for example, i_2) and for any contract given in Figure 3.2.

7.7.6 Interpretation of our approach in terms of generalized localized marginal costs

The short-run marginal cost (SRMC) signal as defined for static optimization purposes in Chapter 3 is a strictly static concept that is useful in interpreting operating cost of generation-based systems control in response to known (scheduled) system inputs. As such, it is not directly useful for allocating control cost for meeting performance over mid- and long-term horizons in response to either short-term scheduled transactions or noncomplying transactions.

It is straightforward to generalize this concept to processes associated with mid-term and long-term dynamic performance according to the criteria given in Chapter 3. Thus, the term $\frac{\partial J_{T_{contract}}^{tertiary}}{\partial P_{Gi}}[KT_t](P_{Gi}[KT_t] - P_{Gi}[(K-1)T_t])$ is the change in total tertiary-level cost borne by control $P_{Gi}^{tert}[KT_t]$ in response to the contract needs as a technical action $P_{Gi}[KT_t]$ is taken at instant KT_t. The total (long-term incremental) cost during the entire contract duration is simply the algebraic sum of these over time KT_t.

Only under the assumption that during the entire contract duration the system is at an economic equilibrium, the term $\frac{\partial J_{T_{contract}}^{tertiary}}{\partial P_{Gi}}[KT_t](P_{Gi}[KT_t] - P_{Gi}[(K-1)T_t])$ is identical for all SMPs in response to $T_{contract}$. (This would be equal to a time-averaged cost over $T_{contract}$ based on localized marginal cost computations at each KT_t.)

An advantage of this measure over the static localized marginal cost lies in the possibility of incorporating not only static security criteria but also quality of system frequency and voltage over relevant time horizons (temporal component of energy value). Also, a more efficient fixed-price future pricing mechanism is possible than with any other method proposed that does not take the temporal component into consideration.

An important conjecture: Prices set according to (7.48) and (7.49) for any asynchronously evolving economic point-to-point contract of an arbitrary duration and obligation to serve will provide a means for market clearing (i.e., moving from an arbitrary market point to an economic equilibrium). The proof for this is quite involved mathematically, but it is based on the fundamentally simple arguments underlying the information balance method (IBM) [39, 40].

7.7.7 From cost-based to value-based future pricing: incentives for high-quality systems control services

At an economic equilibrium without any externalities (solely supply/demand), the results of centralized (tertiary-level) decision making according to a single performance criterion (7.42) are identical to the competitive supply/demand solutions: All market participants operate at an equal localized marginal cost. This result is not directly useful for establishing fair and reasonable pricing for transmission and ancillary services needed to accommodate competitive supply/demand. Our basic approach is to allow for value-based pricing for these services with acceptable upper limits (caps).

Fair and reasonable pricing with caps should be based on the following. Viewing SMPs as players in the competitive supply/demand market, we make the SM cooperate with CM. As long as CMPs are operating at conditions in which their marginal cost is lower, they will be making profit. This will be facilitated by means of flexible time-varying ancillary services, which lead to the most efficient operation of the entire system. By cooperating with the supply/demand market and helping CMPs to make a profit, SMPs will be driving the system to its most efficient operation. This is the basis for our proposed value-based pricing for ancillary services.

In a competitive environment it will be necessary to estimate at the tertiary level the utility (value) function of ancillary services to each CMP. This idea is common to many other competitive markets and could easily be used at the tertiary level, assuming state estimation tools are available. Typical utility functions are monotone functions of quantity (supply, demand); simply by estimating the trend of power input to market player i, one could over time generate these value functions with various degrees of accuracy. (This again calls for systematic accounting of what is already available at the tertiary level.) It would be conceptually fair to charge on the basis of these value-based measures (7.48) and (7.49). There exists a potential for profit making by SMPs while still clearing the supply/demand market.

One should bear in mind the conceptual differences between (1) providing systems control at each subsystem level and charges at this level, independently from the rest of the system; and (2) providing coordinated, hierarchically structured management of performance objectives at each subsystem level as well as at the interconnected system level, and the corresponding coordinated pricing for generation-based systems control from the highest level. In analyzing performance of specific industry structures under present consideration, such as "poolco" [27], bilateral [115], and multilateral [35, 59], in terms of their performance relative to ideal efficiency within operating constraints, it is important to be specific about this division of responsibility and a potential discrepancy between the technical signals and their value allocated under a specific industry structure. This is briefly analyzed next.

All three industry structure scenarios are particular examples of the two-level operations framework described earlier. Consider a nested hierarchical structure, consisting of administratively divided subsystems yet electrically interconnected. Within each subsystem there may be numerous nested CMPs.

7.7.8 Revisiting the "poolco" structure

In this industry structure, all pricing contracts are set at the tertiary level, the participants are not scheduled for power quantities, but the price is computed at the interconnected system level by means of a static optimization algorithm, an OPF at best. All market participants are given a *bundled* price of a hypothetical bilateral deal and the cost of *static* systems control services. This is a particular case of all participants being in the set N_A, with the set N_C being an empty set. This means that all participants are equally responsible for the risks/opportunities taken when projecting into the future the values defined in (7.48) and (7.49) over the contractual time. These figures are bound to be different at the end of the contract. If this option is adopted, then according to (7.52) and (7.53), *all* auxiliary services divide the additional loss (or profit) according to these formulae. Another important observation in the context of the approach described in this text is that under this structure there would be no guarantee that performance criteria over the mid- and long-term horizons would be met.

7.7.9 Revisiting the bilateral structure

Future pricing contracts are set between any two players (two CMPs, subsystem and one CMP within a different subsystem, two subsystems) independently of the situation on the interconnected system. While it may appear to be harder to account for systems control services in this structure, this is actually quite straightforward with our approach.

Two qualitatively different scenarios are possible here. First, the player setting the contract is responsible for not disturbing the rest of the system, in the sense that the *net* interface flow between the player and the rest of the system is unchanged. This is achieved solely by the control actions of the specific player, and it means that no additional bookkeeping between the system and this player is needed. A generalized AGC scheme can do this.

Second, the player engaged in an economic contract does not take responsibility for regulating the flows to the precontract values. In this case, it is necessary to account for the economic value of the SM to this transaction, defined as (7.48) and (7.49), and at the contract termination charge the player engaged in the economic transaction for the difference. The bookkeeping process for this may become a nightmare, unless done with the minimum information needed. The formulae basically relate *only* individual net flows to tertiary-level controls.[14]

7.8 Summary

This chapter is motivated by numerous debates over the value of control services in an increasingly competitive energy environment. Prior to proposing a specific definition for this quantity, we review several relevant issues. The question of optimal performance in a competitive environment is analyzed first. We conclude that the objectives of energy management could become quite different than in the present utility structure. This will require new ways for performance-based valuing of services to different participants in the energy market. Particularly complex to quantify are the transaction-specific value of control services and systemwide controls in a competitive environment. This problem is often discussed in the context of issues such as parallel path (loop) flow problems and the comparability of service [113].

In spite of seemingly very complex issues underlying a potential framework for valuing actions of different participants under open access, it is proposed that fairly straightforward generalizations of methods routinely used by many utilities at present would be sufficient for adequate energy management in a competitive utility business. This is illustrated in the context of optimal real power scheduling subject to frequency regulation and the sufficient information structure needed. It is shown that a systematic unbundling of an AGC-like information structure could be a very effective way of quantifying a technically sound, performance-based value of control services and supporting controls. By means of a generalized AGC-like information structure, control-driven models are derived that define system changes over long time horizons in response to slowly varying loads and open-access-related activities. *Only variables directly relevant for the interactions among the subsystems are used, resulting in extremely manageable models.* While the AGC is an inherently decentralized information scheme, it is proposed here that the same generalized information structure has to be used for the effective centralized control coordination needed for efficient energy management.

Regional electric power management structures of the future are likely to cover a wide span of control and management frameworks. In a centralized

---

[14] The discussion is limited to real power regulation only. Similar reasoning is applicable to generation-based reactive power regulation.

arrangement it would be possible by means of the proposed reduced information structure to achieve a systemwide optimum. Optimality issues with a decentralized implementation need further study. A mathematical derivation quantifying these differences is given in Section 7.4. The formulae can be used as on-line measures of "goodness" of the industry structure adopted, with the best[15] implementation having zero difference. A performance-based valuing of control services is possible in both cases by means of the simple information structure proposed. Inherent limitations and potential risks of an entirely decentralized energy management scheme, which would mimic economic markets, are analyzed.

Using the proposed information structure, a general definition of a control service value is introduced. The definition applies uniformly to all types of subsystems in any industry structure. It provides a good measure to different participants (subsystems) and is suggested as the basis for performance-based pricing. A systematic application of concepts proposed in this chapter provides direct answers to the problem of parallel path (loop) flows and comparability issues.

The concept introduced in this chapter allows for multiple players as well as for nested systems, such as NUGs within a utility, which can also be interconnected to other utilities. The model is sufficiently general. Illustration of these general results is given in the context of both real and reactive power controls.

This chapter revisits the use of localized marginal cost-based pricing. It is shown how one can unbundle specific services, such as frequency and voltage support, and provide quantification of their total value to each market participant. The approach allows for a nonconformal pricing [21] of generation-based ancillary services. The contribution allocation is accounted for over various time horizons relevant for these services. The approach is not a marked departure from the way pools and utilities operate at present. The contribution is primarily related to nondiscriminatory allocation of charges for these services.

It is shown that localized marginal cost-based pricing is a particular example of our proposed value-based future pricing mechanism in which the temporal aspect is not essential, provided only extremely short-term, firm energy contracts are envisioned. Otherwise, for longer, uncertain contracts it is hard to use only the spatial component in pricing. Moreover, the noncompliance component must be accounted for under any industry structure.

Notions like very much needed *long-term incremental costs and economic values* over the periods of long, fixed-price contracts are simply defined by introducing concepts such as (7.48) and (7.49). The proposed approach links operations planning and pricing in a smooth manner. It is straightforward to implement in present pools.

---

[15]This is in the context of the optimal charging mechanism as defined in [117].

CHAPTER 8
NETWORK-BASED SYSTEM REGULATION

8.1 Engineering issues and opportunities in operating power transmission grids of the future

This chapter is concerned with the pivotal role of the transmission grid in achieving energy efficiency over interconnected power systems. The transmission grid is taking on greater importance in response to competition and deregulation in the power industry. We first briefly review its role and the methods by which it has been compensated and controlled in the past. In the future, it is likely that the grid will be called upon to accommodate large energy transfers over far electrical distances, maximize power flow over contractual paths while minimizing flow changes in the rest of the system, and accommodate additional transfer needs caused by an increased presence of independent power producers (IPPs) on the system. While much attention has been given to the regulatory aspects of these changes, it is also necessary to review how the changes will affect engineering issues in planning and operating the transmission grid.

This chapter focuses on the engineering aspects of the transmission grid. It reviews fundamental planning and operating questions relevant to the transmission grid, as seen by the authors. Its emphasis is on network control concepts considered essential in planning new compensation of the transmission grid, and the adaptation of the grid to dynamic and uncertain changes in generation and load inputs to the grid.

8.2 Recent changes affecting the transmission grid and their relation to the basic engineering issues

The primary role of a transmission grid[1] in a normal operating mode is one of a firmly connected electric network in which transmission lines are freely utilized to transmit power from where it can be most economically generated to where it is needed at the moment. The transmission grid typically connects all the generating stations of the system, which range from nuclear to renewable plants. The transmission system governs the power exchange by handling very large blocks of energy. However, the grid is subject to inherent power transfer limits. It is because of this that the solution to any transmission problem represents a compromise between engineering needs and economic feasibility.

Consider now the changes seen in input patterns to a transmission grid that are utility-dependent. On some systems dynamic inputs are seen on the generation side, whereas the load side is not directly regulated and is only subject to anticipated load uncertainties. In contrast, other systems, which encourage demand-side management and load regulation, have dynamic patterns at the load buses. On the U.S. transmission grid specifically, many of these changes are triggered by government (federal or state) regulations that tend to encourage the presence of small IPPs on the generation side as well as demand-side management on the load side within any given utility. This generally results in more dispersed generation throughout the system instead of large electrically distant sources of energy. In most cases small dispersed generation sources are not utility-owned and therefore not directly dispatched and controlled by the utility to whose transmission grid they are electrically connected.

The transmission grid connecting different utilities within a large power pool is also subject to more dynamic inputs than in the past. Again, on the U.S. power grid, these changes are government-regulated at present, primarily by imposing an open access or wheeling mode on the pool. Even for utilities that are not subject to new government regulations of the sort present on the U.S. system, one can see a natural tendency to operate the system in the economically most effective way, which often requires the transmission grid to operate far from its nominal design conditions. In both cases, the net effect on the transmission grid is similar in terms of having more dynamic and demanding inputs than in the past.

While the regulatory changes are well documented throughout the literature, there seems to be a dearth of power engineering publications that identify and quantify the engineering effects of these changes on the transmission grid. Finding systematic ways to control the grid under these unusual

---
[1]The material in Sections 8.2–8.6 is a reprint from [23].

Chapter 8

and dynamic inputs is a necessary task. The material in this chapter is written in an attempt to provide an engineering assessment of the changed requirements for the grid and the basic technical challenges posed by these changes.

First, one can argue that the grid should be operated such that its main role of transferring power from where it can be most economically produced to where it is needed at the moment is carefully preserved. For energy consumersnot responsive to prices, nothing should be changed with regard to a utility's basic obligation to provide high-quality steady-state voltages, no matter what else is happening on the system. Energy consumers participating in demand-side management, as well as the IPPs, are seen by a transmission grid in a similar way. As long as they are present in small numbers, they can be treated as uncertain loads and managed by a transmission grid in traditional ways [122]. However, if these time-varying inputs approach a level comparable to utility-owned generation, the energy management, including both planning and operation of the grid, will require significant hardware and regulation enhancement to meet the same basic task while accommodating widely varying inputs to the grid. Typical technical requirements triggered by these changes enhance the need to increase large energy transfers over prespecified paths while minimizing flow changes through the rest of the system [123]. Although energy management cannot be done in an optimal way by separating the role of system inputs from the role of a transmission grid, it is shown here that there are a number of basic planning and operating questions that can be uniquely associated with the transmission grid. This is particularly important for systems where ownership of the grid is separated from ownership of generation [124].

While the engineering questions basic to transmission grid management are generic to both single- and multi-utility environments, this chapter considers only a single-utility setting.

The problems are grouped into basic planning questions and basic operating problems. Basic planning questions could be further grouped into

- Quantification of the limits on power transfer with the existing hardware on the transmission grid.

- Defining the characteristics and most effective placement of new compensation equipment that could relax existing transfer limits without violating the prespecified thermal, voltage, and stability limits.

The basic operating questions are primarily related to the ways in which compensation devices on the grid are controlled. The hardest question is how to design effective systemwide controls and use them in a robust way.

In order to systematically introduce the operating problems, we describe transmission grid operation by

- Reviewing ways in which elements of the transmission grid are regulated at present, identifying otherwise acceptable operating regions in which the present control logic may not work effectively, and proposing improved logic to enlarge regions of effective regulation of the transmission grid.

- Defining opportunities offered by the new semiconductor-based controls and developing a framework for their best control.

In this chapter, the intent is to provide a global outlook on the basic technical problems evolving on a transmission grid when it is viewed as a central player in energy management. Because of space constraints, some topics are given minimal discussion, but many references are cited for further reading.

8.3 A brief review of the present principles for regulating a transmission grid and the power system

Let us consider the simplest possible transmission grid topology of one line connecting a single generator to a single load (Figure 8.1). To emphasize the technical aspects specific to this grid, we view changes at the generator and load ports simply as input changes to the transmission grid. (Note that the thinking here is quite unconventional and different from that typically practiced, in which much emphasis is put on generation while the transmission grid takes on a passive, secondary role.) As described earlier in this text, inputs to the transmission grid are regulated via local generator controls, such as an excitation system intended to maintain its steady-state terminal voltage V_1 within a preset threshold of V_1^{set}, and a governor control regulating its real power input into the grid P_g at P_g^{set}. Loads are typically assumed static and are characterized by their real and reactive power inputs into the transmission grid.

Note that except for the natural dynamics of generators, the typical time scale separation of different phenomena evolving on the transmission grid is hierarchical and control-imposed. The generator local control is the fastest. The slowest is the systemwide regulation of terminal voltage set points at the generators and selected loads in response to typically slow load input variations. This level is managed from the control center of each utility. An important question here is that of determining how the compensation of the grid should be managed, i.e., planned and operated, to meet the desired set values at the terminals to the transmission grid. Roughly speaking, this problem is one of providing for the right on-line compensation of the transmission grid and for the control of the compensating devices so that the optimal load

Chapter 8   199

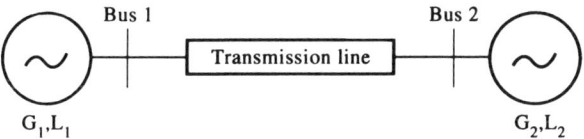

Figure 8.1: The simplest power grid topology

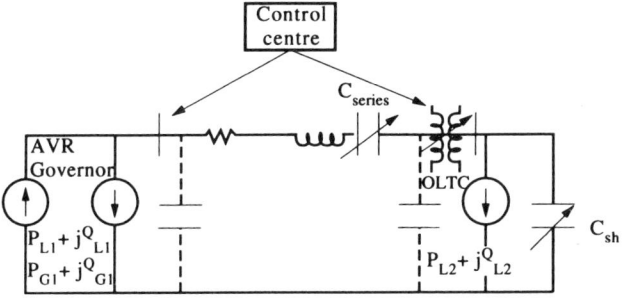

Figure 8.2: Hierarchical, control-imposed time scale separation

flow as defined at the control center level is achieved as the inputs to the grid (generation or demand) vary.

The main compensating devices at present available on a typical transmission grid are reactive in nature, with the exception of DC transmission, and are slowly regulated via mechanical switching. The most typical compensation is provided by shunt capacitors/reactors and automatic tap changing transformers, and more recently, by series capacitors. As portions of these devices get connected to the grid in discrete amounts, properties of the transmission grid change and can be simply quantified by looking at changes of its reactance (or impedance) matrix [126]. Typically these compensating devices are installed at the particularly vulnerable locations of the grid in order to directly regulate load voltages. Their regulation could be manual or automated. Some of these compensation devices are schematically represented in Figure 8.2.

8.4 The basic planning problem on a transmission grid

The basic planning problem is one of designing a transmission grid needed to accommodate power transfer between any two buses on the grid, given the anticipated bounds on inputs to the grid (both generation and loads). The grid is typically designed to meet this goal for nominal loads and for anticipated variations. In the past, only fine-tuning of the transmission grid parameters has been required, since the loads were quite certain and very slowly changing. Additional compensation of the grid has been typically achieved by switching reactive devices in response to load changes. However, in light of more frequent substantial deviations in inputs from the values for which the grid was originally designed, additional planning of significant amounts of reactive compensation may be needed to make the desired transfer between two given buses on the grid feasible, and possibly optimal. (It is generally recognized that adding new transmission lines is not a realistic option, but their enhancement by means of various compensation devices is.)

This planning task (adding new reactive compensation and making it optimal) is fairly well understood for the simplest transmission grid topology of Figure 8.1. Because of this, it is reviewed here first for this topology and then generalized to planning problems for grids of general topology.

The feasibility of a desired power transfer is directly related to the notion of a steady-state transfer limit. Its simplest interpretation is as a conventional problem of real power transfer, illustrated in Figure 8.3, under the assumption that the reactive power/voltage support is such that voltages do not change very much. The maximum feasible real power transfer on the given transmission line is parametrically dependent on the sending and receiving voltages at the terminals to the line, as well as on the reactance of the transmission line as

$$P_2^{max} = \frac{V_1^{set} V_2^{set}}{X_{12}} \qquad (8.1)$$

The input regulation on the generator side directly affects the value of the V_1^{set}. However, the other parameters X_{12} and V_2^{set} are directly affected by different means of transmission line compensation. V_2^{set} can be maintained by adding shunt compensation $\frac{1}{j\omega C_{sh}}$, which is switched in response to deviations of V_2 from V_2^{set}, and also by the automatic tap changing of a transformer if one is present on the line. Similarly, a series compensation $-j\frac{1}{\omega C_{ser}}$ can be added to reduce X_{12}, and consequently increase P_2^{max}. The proper amount of any of these compensating devices will achieve the same goal of changing P_2^{max} to its new desired value.

Note that these compensating devices are reactive in nature. While the main purpose of the grid is to deliver real power in an economic way, this

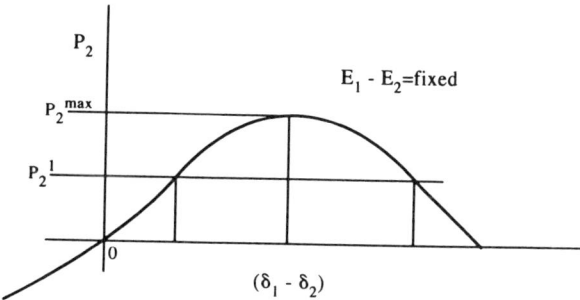

Figure 8.3: Real power transfer limit

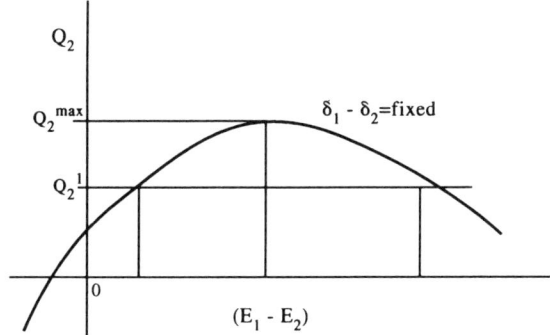

Figure 8.4: Reactive power transfer limit

transfer is greatly affected by reactive power devices.

In light of recently observed voltage problems, it is also important to recognize that the design and planning of the grid has to provide reactive support sufficient to supply a specified reactive load demand Q_2^{max}. Because of this, one must also monitor the reactive-power-related steady-state transfer limit, pictorially represented in Figure 8.4. Here real power transfer, and therefore voltage phase angle difference across the transmission line, is treated as a given parameter. While the most common notion of a steady-state transfer limit is associated with the setting shown in Figure 8.3, compensation of the grid may be more critical in some circumstances with regard to the reactive power transfer limit, shown in Figure 8.4.

In the early literature, this basic steady-state transfer feasibility problem is referred to as the line loadability problem [127, 128]. For an extensive review, we refer the reader to these articles.

The main new aspect of planning with regard to power transfer limits in a deregulated environment is concerned with transmission grids of general topology for which new compensation may be needed to enhance power transfer only between selected buses on the grid. These buses could be defined as a result of particular contractual needs, or in response to the presence of IPPs, or simply caused by the need to achieve the most economic transfer within a traditional setting of a given utility.

A typical approach at present is to run large numbers of scenarios using load flow programs and detect limits of feasible transfer when load flow begins to have convergence problems [129]. Unfortunately, it is hard to associate nonconvergence of the load flow program with the particular transmission lines reaching their power transfer limits. More advanced analytical tools are needed to directly identify transmission paths on a large grid that first reach the steady-state transfer limit for the specified real power input changes at a few buses on the grid. This question is often referred to as the problem of available transmission capacity (ATC) [146]. It is also important to estimate a bound on the increase in real power inputs at the chosen bus(es), so that the line flows remain within the prespecified limits. Results of this type would be very useful for accommodating planned deviations from nominal only in the areas of direct interest. We call the reader's attention to such planning methods introduced in [47].

The second major part of planning a transmission grid to be most adequate for specific changing patterns in power transfer is the question of its optimal compensation, i.e., the question of adding the best compensation for a chosen performance criterion. Keeping in mind that the fundamental function of a transmission grid is to transfer real power according to anticipated patterns, the optimal compensation via reactive devices can be interpreted in a fairly straightforward way only on the simplest transmission grid topology (Figure 8.1). It is not hard to show that the following optimization problems are equivalent on such a topology [62].

- *Problem 1:* Find a reactive compensation that will provide for minimum real power transmission losses for a given real power input at the generator side P_g and load voltage V_2. Its mathematical formulation, with the notation given in Figure 8.5, is

$$\min_{i(t)} \frac{1}{T} \int_0^T R_e i^2(t) \, dt \tag{8.2}$$

subject to

$$\frac{1}{T} \int_0^T i(t) u(t) \, dt = P_g \tag{8.3}$$

and

$$V_2^2 = \frac{1}{T} \int_0^T u^2(t) \, dt \tag{8.4}$$

- *Problem 2:* Find a reactive compensation that minimizes reactive power input into the transmission grid, as seen at the source of energy (generator) terminals, for specified P_g and V_2.

- *Problem 3:* Find a reactive compensation that maximizes P_2 for specified P_g and V_2.

All three problems are solvable as a simple calculus of variations problem [130], and have the identical solution:

$$i^*(t) = \frac{P_g}{V_2^2} u(t) \tag{8.5}$$

The typical compensation needed to change $i(t)$ to its optimal waveform $i^*(t)$ can be either shunt or series. For the example given in Figure 8.1 and the load characterized by its impedance $R_l + jX_l$, the optimal series compensation is

$$R_{ser} - jX_{ser} = \frac{X^2}{R_e} - j\frac{1}{L\omega^2} \tag{8.6}$$

and the optimal shunt compensation is

$$C_{sh} = \frac{X}{R_e^2 + X^2} \tag{8.7}$$

This solution can be directly related to the well-known solution of maximum power transfer in general electric networks [131, 132].

It is important to recognize, however, that a power transmission grid of typical design cannot be operated close to the optimum defined via (8.5). This is because at this global optimum the load voltage V_2 would typically become excessively high. This can be seen from the simple relation between load and generator voltage

$$V_2 = \frac{2Re}{\sqrt{R_e^2 + X_e^2}} V_1 \tag{8.8}$$

Additional constraints are needed to guarantee that the voltage remains within the acceptable, prespecified limits during any attempts to achieve unusually high real power transfers. One of the first formulations of the constrained maximum power transfer was presented in [133] and is derived under the additional constraint of a constant load power factor. It has been further shown that this constrained (local) optimum power transfer occurs when the magnitudes of line and load impedances become equal. It is also

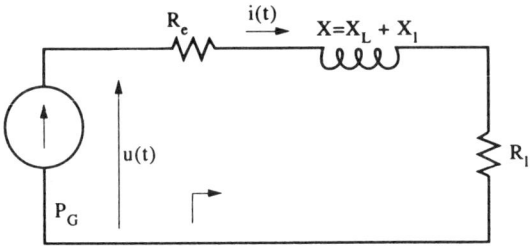

Figure 8.5: Optimal series and shunt compensation

simple to show that this local optimum results in the maximum possible power transfer given as [17]

$$p_m = \frac{1/2}{\sqrt{(1+r)^2(1+q)^2 + (r+q)}} \qquad (8.9)$$

The corresponding load voltage is

$$v_m = \sqrt{1/2 - (r+q)p_m} \qquad (8.10)$$

where

$$r = Re/Xe, \quad q = Pl/Ql, \quad v = \frac{V_2}{V_1}, \quad p = \frac{X_e P_l}{V_1^2} \qquad (8.11)$$

A closer look at the voltage limits as a function of compensation reveals a fairly small range of effective reactive compensation within which V_2 remains below typical operating limits.

Because of the emphasis of this chapter on reactive compensation, another powerful compensation for the grid is not analyzed here, i.e., a phase-shifting transformer. It would be interesting, however, to establish conditions on constrained (local) power transfer optimization by allowing for a phase-shifting transformer as a compensating tool on the grid. It is likely that this type of compensation would provide for larger power transfers while maintaining the load voltage below the prespecified limit.

The problem of optimum power transfer via a transmission grid of general topology is much more complicated than that via a single transmission line. First, it is hard to directly relate total real power loss minimization to the optimal real power transfer between two chosen buses of a general transmission grid. Furthermore, Problems 1, 2, and 3 defined above for the simple grid do not have the same solution for a general topology, and therefore are not equivalent. This brings back the basic question of the most meaningful performance criterion to associate with optimizing real power transfers. It should be noted here that while the standard optimal power flow is often viewed as a tool for planning reactive power compensation of general transmission grids, its results are often not very practical because they cannot be directly interpreted in terms of flow compensation on specific transmission paths of greatest interest.

To appreciate the issues discussed here, the reader is alerted to some very interesting and potentially useful definitions of transfer efficiency between two chosen buses on the general DC electric network [134, 135]. It is our assessment that similar developments will be needed for AC power transfer in order to develop a basic understanding of optimal real power transfer on a general transmission grid and to use it for the design of optimal reactive compensation.

8.5 Operating problems using mechanically switched reactive devices

In contrast to the planning problems of adding new software, the operating problems are directly associated with the control of individual reactive devices on the grid and their coordination so that the desired steady-state voltages are maintained. Most of the reactive devices at present available on a typical transmission grid are switched in anticipation of a particular load pattern. Others are automated to respond to local voltage changes at buses they are expected to regulate. Most of the automated switching is mechanical and designed with an intentional delay to separate the process of steady-state voltage regulation on the grid from the natural system dynamics and the fast continuous regulation of generation inputs to the grid.[2]

Most typical automated switching of the reactive devices can be illustrated for the case of an automatic tap changer of a transformer [136, 137]. The tap position of a transformer connecting buses i and j and directly responding to deviations in V_j from V_j^{set}, $a_{ij}(k+1)$ at the $(k+1)$st regulation step, is

---

[2]The interplay between generation-based systems control over mid- and long-term ranges and the mechanical switching of grid devices must be studied more carefully.

related to the tap position $a_{ij}(k)$ at the previous step as

$$a_{ij}(k+1) = a_{ij}(k) + d_j f(V_j - V_j^{set}) \tag{8.12}$$

where the switching logic is of relay type

$$f(V_j - V_j^{set}) = \begin{cases} +1, & (V_j - V_j^{set}) > \delta V_j \\ 0, & (V_j - V_j^{set}) < \delta V_j \\ -1, & (V_j - V_j^{set}) < -\delta V_j \end{cases} \tag{8.13}$$

and d_j is a chosen regulation step. Much is known by now about sufficient conditions for the control logic (8.12) and (8.13) to regulate voltages to within a desired threshold around V_j^{set} [73, 137]. Moreover, some voltage-related operating problems in the past are directly attributable to malfunctioning of these automatic tap changers. The fear of their repeated malfunctioning has led to operating policies of locking or disabling their automated response in low voltage operating regions. This situation calls for more work on improving the present control logic of automatic tap changers, which would guarantee their reliable performance within the widest possible range of voltage deviations, instead of disabling them when they may be most useful. There have been sporadic results reporting potential use of improved, more adaptive logic in switching automatic tap changers [73, 138]. In general, such improvements would enable better operation of available reactive devices on the grid.

Note that we do not formulate mathematically the notion of "optimal" in an operating environment, because such formulation is simply not available and not generally agreed upon at this time. Therefore, "optimal" should be understood in a qualitative setting of operating available devices in such a way that they allow for regulation of the widest possible deviations from nominal over a specified time interval of interest.

It is our conjecture, however, that a reactive device is operated closer to its optimum (defined in a qualitative sense) if the integral of the square of its reactive power input to the grid over the time interval of interest is minimized. There has been some evidence for the correctness of this conjecture in the context of interpreting improved nonlinear excitation systems [139, 140]. Provided this conjecture proves true in general when comparing any two specific switching laws, this would indicate that the generalization of Problem 2 formulated in the previous section is most meaningful for the operating environment as well.

With such unified performance criteria of the type directly expressed in terms of reactive stored energy, it would be further possible to quantify and mutually compare the "goodness" of specific control schemes in the operating environment. This would begin to assess, then, the relative value of particular controls, which is often missed while attempting to suggest better controls on the system. Fundamental work is needed in this direction to define similar notions on a transmission grid of general topology.

It is interesting to further interpret the effects of mechanical switching in terms of parametric changes of the transmission grid characterization [126]. From the operating viewpoint, it is important to understand that an automatic tap changer installed between buses i and j to directly regulate load voltage V_j will operate as desired following the standard switching logic (8.12) and (8.13) as long as the parameters of the modulated transmission line behave as parameters of a normal mechanical spring, i.e., as a spring that stretches more when more force is applied to it. This can be further interpreted in terms of the operating region to the left of Q_2^{max} in Figure 8.4 [73]. The characterization of a transmission grid in terms of its admittance matrix properties reveals that, as long as the admittance matrix remains positive-definite, the grid will behave as a normal spring, i.e., any increase in load demand will reduce load voltage, and vice versa.

Similar analysis of other types of reactive devices regulated according to relay type logic (8.12) for automatic tap changers leads to the same conclusions. For example, given parameters of an uncompensated transmission grid in terms of its reactance matrix X, there exists a bound on the effective shunt compensation C_{ii} at each bus i that, when subjected to automated switching logic of type (8.12), would regulate the voltage in the desired direction. This bound is defined as [107]

$$C_{ii}(k) < \frac{1}{X_{ii}(k)} \qquad (8.14)$$

where $X_{ii}(k)$ is the Thevenin reactance as seen from the load port i at the switching step k. For shunt capacitors larger than specified by (8.14) it may be necessary to improve the switching logic in order to regulate voltages within their largest possible regions.

Much work is still needed to develop an adaptive regulation of the mechanically switched reactive devices at present available on the transmission grid in order to achieve robust voltage regulation to their values set at the control center level. An additional requirement needs to be observed whenever possible. That is, automated switching is at present practiced in a local, decentralized way; given a local desired voltage, the switching of a specific reactive device responsible for its regulation is done using only the information about this voltage. This decentralization should be preserved whenever possible while proposing new adaptive switching techniques.

It is unlikely that the system will often enter operating regions with unusual problems calling for adaptive regulation, provided intelligent planning of reactive devices is used for their design according to criteria discussed in the previous section. This is an unproven conjecture, which can only be supported by extensive simulation at present. For example, if hard operating limits on voltage are observed at the planning stage, it is unlikely that regulation will be needed for voltages below those values. It has been our persistent observation in simulating many realistic power systems that the voltage value corresponding to Q_2^{max} in Figure 8.4 is never lower than the hard operating limit on minimum voltage. Of course, some exceptions are possible, but we have not found them.

8.6 Opportunities and problems presented by very fast regulation of the transmission grid: FACTS trends

Up to now, we have not considered possible transient stability problems in attempting to operate the grid in a most economical way. This is because the slow mechanical switching discussed so far is only capable of regulating steady-state voltages. In the past, transient stabilization has been primarily a responsibility of fast continuous input regulation of generators. It has been only very recently that we are presented with qualitatively new hardware opportunities to stabilize system dynamics also by the very fast regulation of a transmission grid. A variety of newly developed power devices based on electronic/microprocessor switching of standard reactive compensation devices recognized under the acronym FACTS (flexible AC transmission systems) can be viewed as a class of devices that could potentially drive a transmission grid response in a time domain that is impossible to achieve by slow mechanically switched reactive devices. Much work is needed to define opportunities offered by the fast semiconductor-based controls of standard reactive devices on a transmission grid. What is new is the speed at which the power-electronically switched reactive devices respond to dynamic changes on the system. This presents us with new locations on interconnected power systems where one could respond as is now possible at the generation inputs to the grid regulated via fast excitation. In general, an entirely new control framework for the "best" regulation and systemwide coordination of these devices is needed. It can be conjectured that on the simplest transmission grid topology (Figure 8.1) with one candidate FACTS device, the "best" control logic will be the one that leads to the minimum integral of the reactive power output over the time interval of interest, all other conditions being the same. This can be shown by simply replacing the excitation system of a generator by a static Var compensator in order to fast regulate the local voltage and applying already known conclusions from [139]. It is interesting to view a static Var compensator simply as a device that directly modulates the parameters of the transmission grid by very fast changes of its effective shunt capacitance/reactance. Similar statements can be made about the fast modulation of series capacitors. With this in mind, one could propose measures of performance of FACTS devices in terms of the integral changes of stored reactive energy in the regulated device over the interval of interest.

This is straightforward for the simplest transmission grid topology modulated by a single FACTS device. Its importance is in providing ways of comparing performance of different controls on FACTS devices, and therefore quantifying their value to the system. This thinking is analogous to analyzing the static switching of reactive devices, with the exception that FACTS switching has a potential of helping true dynamic problems on the system, while the mechanical switching regulates the system within steady-state voltage limits.

The general lack of understanding of how to systematically stabilize systemwide dynamics via the fast local modulation of the reactive devices on the grid should not be surprising. Our knowledge of systematic techniques for any local stabilization of power system dynamics, even via more conventional devices such as excitation and governor systems, is rather limited [24]. The state of the art is such that very little can be claimed in terms of robust local stabilization of the type arising in power systems, without requiring an excessive number of control locations [141, 142]. Without fundamental research in this area very little use will be made with full confidence of the real opportunities offered by FACTS devices. For the time being we only have limited examples, entirely based on simulation, which demonstrate that fast regulation of reactive compensation on a transmission grid could be very useful in the future [19, 20]. Because of this, there may exist an immediate danger of uncoordinated systemwide fast regulation via FACTS devices that could become detrimental to system integrity under certain operating conditions.

Work is needed to establish nonconservative measures of stable regions achievable via fast FACTS regulation of the grid. Only a systematic approach to this goal will contribute to the long-term benefits that can be achieved by the opportunities offered by FACTS technologies. Generalizations of open questions and partial answers presented in the context of the state of the art in mechanical switching have to be made for the fast dynamic phenomena of interest when regulating via FACTS devices. Otherwise, assuming local fast regulation of these devices without much knowledge about their systemwide effects could lead to more harm than good. One possibility for use of FACTS devices is to allow the system to operate in previously unused regions and still guarantee stable robust regulation. When doing this, one must be extremely careful, since there will not be much available operating experience.

8.7 Direct flow control via FACTS devices

As discussed in Chapter 4, inter-area dynamics represent singular modes of the system under the present local control structure. Because of small dampings, these singular modes often occur as oscillatory. Under weak interconnections, the oscillations become slow and persistent [79, 80]. As shown before, the structural singularities cannot be removed by any design under the present local control structure. We propose a new control scheme to remove the structural singularities by directly regulating tie-line flows using the fast power-electronically switched controllers now being developed and tested. These controllers are often referred to as FACTS devices [19, 20]. The idea of the proposed new control scheme is to change the dynamic characteristics of the inter-area modes by a feedback control so that the oscillatory behavior of the inter-area modes becomes exponential and settles in a desired time constant. First, the case when all tie lines are equipped with FACTS control devices is studied. Detailed control design procedures are given. Next, the case where only a limited number of tie lines are equipped with the controllers is briefly discussed.

8.7.1 All tie lines directly controlled

As proposed in the general theoretical setting in Chapter 4, tie-line flows act as direct control inputs to the regional dynamics using FACTS devices. The additional control inputs are used to remove the structural singularity under the present local controls. Consider any administratively divided region within an interconnected system. If all components of the equivalent tie-line flows F_e are assumed to be the direct control variables with FACTS devices, model (5.56) can be viewed from a control design viewpoint as being of the form

$$\dot{z} = -l^T u + d \tag{8.15}$$

where $u = \dot{F}_e$ represents the additional control variables for the administrative region, l^T is the participation factor vector, and $d = l^T D_P \dot{P}_L$ is the disturbance caused by the typically not measurable load variations $\dot{P}_L$. The equivalent tie-line flows F_e defined in (5.37) as a combination of flows into area generator nodes F_G and into the load nodes F_L are the new control variables to be designed according to specifications of the inter-area dynamics. If the inter-area dynamics are to meet particular response characteristics, including elimination of slow, persistent oscillations, specific flows F_G or F_L will need to be controlled. Notice that model (8.15) of inter-area dynamics can be seen as entirely control/disturbance driven. In the ideal case when all tie lines are equipped with the additional control hardware capable of directly regulating real power flows, each area could directly regulate its interaction variable z, responsible for interactions with the neighboring systems, by simply regulating it to the scheduled value z^{ref}, according to the general form

$$u = G_p e + G_i \int e \, dt + G_d \dot{e} \tag{8.16}$$

where $e = z - z^{ref}$ is the error of the interaction variables and G_p, G_i, and G_d are the gains corresponding to the proportional, integral, and differential controls. It can be seen from (5.37) that the equivalent tie-line flows F_e corresponding to each specific area can be achieved by a variety of combinations of individual tie-line flows into the boundary generators, F_G, and into the boundary loads F_L. Eq. (5.37) can be used to decide on the most effective locations of individual controllers, which could achieve F_e needed to stabilize the interaction variables to the scheduled value z^{ref}. Figure 8.6 shows the schematic of such controls on the 5-bus example.

For the purpose of illustration, let us take the simple form of the control with only a proportional control, i.e.,

$$u = G_p(z - z^{ref}) \tag{8.17}$$

Under this control, the inter-area dynamic model in (5.56) becomes

$$\dot{z} = A_z(z - z^{ref}) + d \tag{8.18}$$

where $A_z = -l^T G_p$. Note that there is in general only one singular mode corresponding to each region. Therefore, the inter-area dynamics are in general a scalar system. The control gain vector G_p is chosen such that the scalar A_z is a negative number with sufficiently large magnitude to ensure the settling time. This is done easily because the left eigenvector l^T is roughly the vector with all 1's. The constant number z^{ref} determines the steady-state value of the interaction variable. Since the inter-area variable is basically the total area generation, modulo losses, the number z^{ref} will have a decisive effect on the steady-state total area generation, and the steady-state tie-line flows. It can be determined such that the scheduled tie-line flows are achieved in steady state.

Let us use the same 5-bus example introduced in Figure 5.3. In this

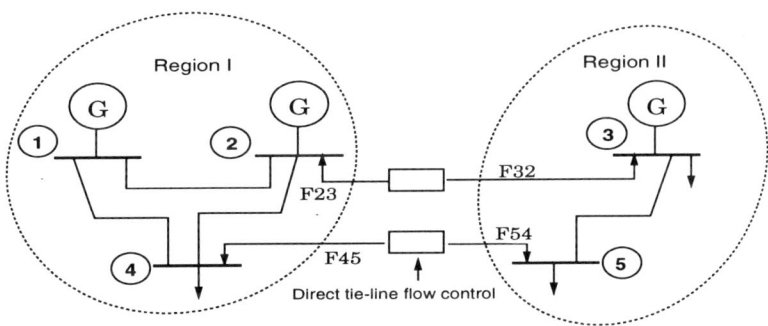

Figure 8.6: Direct tie-line control

example, $F_G^I = [0 \; F_{23}]^T$; $F_G^{II} = F_{32}$; $F_L^I = F_{45}$; $F_L^{II} = F_{54}$. It follows that the interaction variables are measurable directly through measurements of real power generation of the region. In this example, interaction variables are given in the form

$$z^I = l^{I\,T} P_G^I = P_{G1} + 0.9969 P_{G2} \tag{8.19}$$

$$z^{II} = l^{II\,T} P_G^{II} = P_{G3} \tag{8.20}$$

The equivalent flows are calculated from (5.37) as

$$F_e^I = \begin{bmatrix} \frac{\partial P_{G1}}{\partial \delta_4} \\ \frac{\partial P_{G2}}{\partial \delta_4} \end{bmatrix} [\frac{\partial P_{L4}}{\partial \delta_4}]^{-1} F_{45} - \begin{bmatrix} 0 \\ F_{23} \end{bmatrix} \tag{8.21}$$

$$F_e^{II} = [\frac{\partial P_{G3}}{\partial \delta_5}][\frac{\partial P_{L5}}{\partial \delta_5}]^{-1} F_{54} - F_{32} \tag{8.22}$$

The inter-area dynamics of the system explicitly in terms of load disturbances are expressed from (5.56) as

$$\dot{z}^I = l^{I\,T}(\dot{F}_e^I - \begin{bmatrix} \frac{\partial P_{G1}}{\partial \delta_4} \\ \frac{\partial P_{G2}}{\partial \delta_4} \end{bmatrix} [\frac{\partial P_{L4}}{\partial \delta_4}]^{-1} \dot{P}_{L4}) \tag{8.23}$$

and

$$\dot{z}^{II} = l^{II\,T}(\dot{F}_e^{II} - [\frac{\partial P_{G3}}{\partial \delta_5}][\frac{\partial P_{L5}}{\partial \delta_5}]^{-1} \dot{P}_{L5}) \tag{8.24}$$

To illustrate by simulation the effects of direct tie-line flow control on the inter-area dynamics, let us first rewrite the system model (5.42) in terms of voltage phase angle differences. For the same 5-bus example, one has

$$\dot{x} = Ax + N\dot{\Delta} + \dot{d} \tag{8.25}$$

where $\Delta = [\Delta_1 \ \Delta_2]^T$, and Δ_1 is the phase angle difference across the line connecting nodes #2 and #3, Δ_2 the phase angle difference across the line connecting nodes #4 and #5. Matrix N is related to the tie-line reactance. $\dot{d} = [0 - D_P]^T \dot{P}_L$ is the system disturbance due to load variations. It can be shown that only the total (net) power generation of each region is important for inter-area dynamics regulation. Thus, one can assume $\Delta_2 = 0$ for simplicity. Assume further a lossless system, and define $u = -Z_1\dot{\Delta}_1$ as the control signal, where Z_1 is the reactance of the first tie line. Using these notations, the open-loop dynamic model of the region (5.42) becomes

$$\dot{x} = Ax + bu + \dot{d} \tag{8.26}$$

where $b = [0\ 0\ 0\ 1\ 0\ -1]^T$. The open-loop inter-area dynamics take the form

$$\dot{z}^I = u + \dot{P}_4 \tag{8.27}$$

$$\dot{z}^{II} = -u + \dot{P}_5 \tag{8.28}$$

For this small power system the order of the full model (8.26) is three times the order of the local state space x_{LC} augmented by 1, and the dimension of the model representing inter-area dynamics (8.27) and (8.28) is only 2. In general, for more realistic, larger power systems, the orders of the two models differ much more. The proposed tie-line control design uses only the low-order model (8.27) or (8.28).

It is clear that once oscillations in either z^I or z^{II} are suppressed, there will be no inter-area oscillations. This is obvious for the two-area system, since inter-area oscillations are consequences of the power exchange between the two areas. As a result, control design can be done from either side. In this case, since $z^{II} = P_{G3}$, controlling y^{II} needs only one measurement (compared to two for z^I). To illustrate this, assume

$$u = G_p(z^{II} - z^{II\ ref}) \tag{8.29}$$

where $z^{II\ ref}$ is a constant target to be appropriately chosen. The closed-loop dynamics (8.27) and (8.28) for the inter-area dynamics have the form

$$\dot{z}^I = G_p(z^{II} - z^{II\ ref}) + \dot{P}_4 \tag{8.30}$$

$$\dot{z}^{II} = -G_p(z^{II} - y^{II\ ref}) + \dot{P}_5 \tag{8.31}$$

Clearly, the stability requirement for z^{II} dictates $G_p > 0$.

Shown in Figure 8.7 is the case for $z^{II\ ref} = 0.929$ and $G_p = 0.1$, and $z^{II\ ref}$ is chosen to be the steady-state value of net tie-line flow out of area II prior to adding the new controller. G_p is chosen such that the settling time for the inter-area dynamics is roughly 46 seconds. Initial conditions for all states are unity.

As a comparison, Figure 8.8 shows responses of the system when there is no direct tie-line flow control. Clearly, the slow mode corresponding to the inter-area oscillations is eliminated by the proposed control method.

8.7.2 Only a subset of tie lines directly controlled

The fact that the interaction behavior is contained in the inter-area variable z provides some interesting opportunities for the application of the $\mathcal{H}_\infty$ design methodology to the use of FACTS devices for improved transmission grid

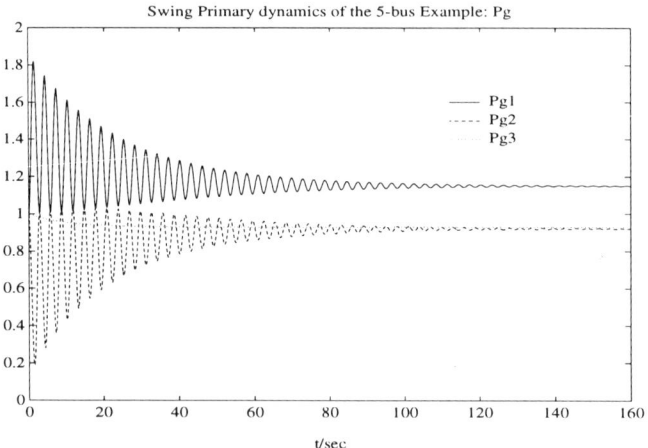

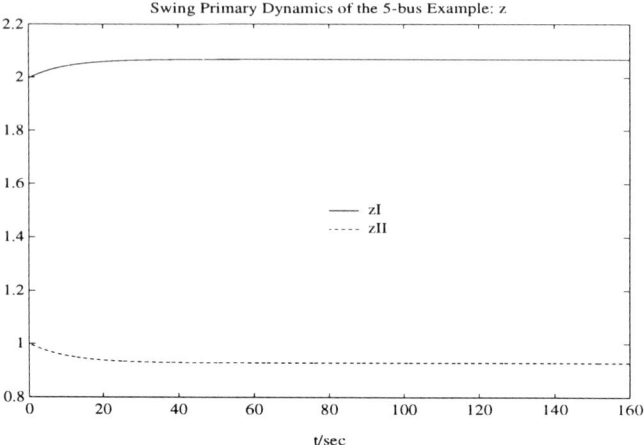

Figure 8.7: System responses after control

Chapter 8

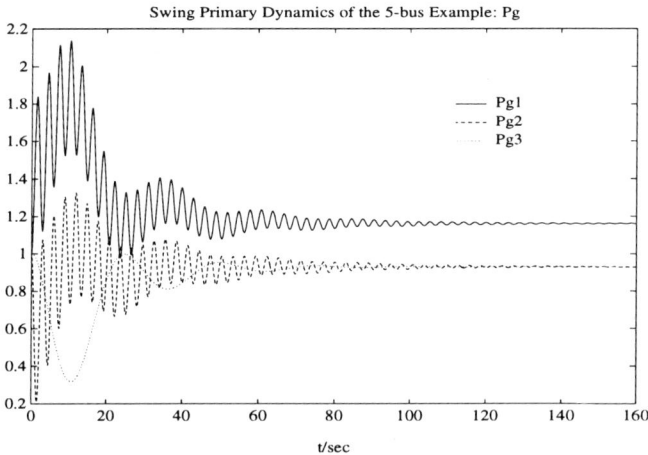

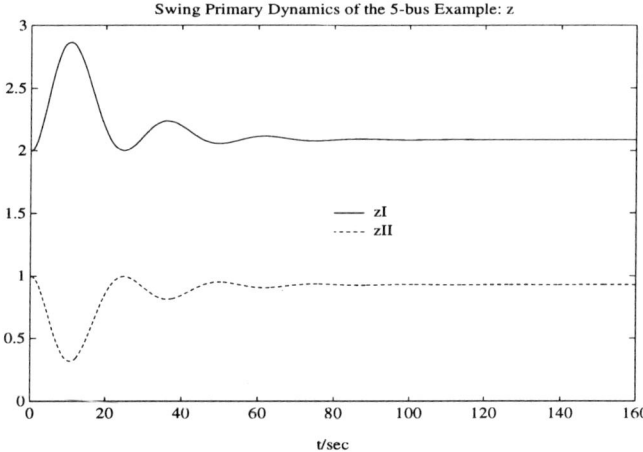

Figure 8.8: System responses before control

response. In order to formulate the problem, we make the following assignments, to follow the notation of [100]. First, we separate the tie-line flows into a controlled group and an uncontrolled one, F_c and F_u, respectively, via a signed permutation matrix U:

$$\begin{bmatrix} F_G \\ F_L \end{bmatrix} = U \begin{bmatrix} F_u \\ F_c \end{bmatrix} \tag{8.32}$$

so that

$$F_e = F_G - D_P F_L = B_c F_c + B_u F_u \tag{8.33}$$

In addition, for the moment it will be assumed that $\dot{P}_L = 0$. We can then write the system model (5.42) as:

$$\dot{x} = Ax + B_c F_c + B_u F_u \tag{8.34}$$

$$z = Lx \tag{8.35}$$

or

$$\dot{x} = Ax + B_c u + B_u w \tag{8.36}$$

in the same notation as [100]. At this point, the controlled output, the disturbance, and the input have been defined adequately for the purposes of the $\mathcal{H}_\infty$ minimization. Appropriate weighting functions $W(j\omega)$ could be defined at the output and input, to tailor the behavior of the resulting controller. Emphasis could be placed on damping a particular inter-area mode, for example, by placing higher weighting at that frequency. Further research is needed to establish accurate solutions to this problem.

In summary, an approach to regulating the inter-area dynamics via direct tie-line flow control using a reduced-order model relevant for the inter-area dynamics is proposed for the first time in [17, 143]. The approach allows for a systematic control design regarding both the best placement of the controllers and their control logic. It is shown that the inter-area dynamics can be regulated using standard designs for low-order linear dynamic systems. An underlying assumption that the new control does not destabilize the intra-area dynamics regulated by generator control needs to be verified as part of the proposed design. The design is based on a reduced decentralized measurement structure at each regional level. The physical devices for controlling the inter-area oscillations are assumed to be capable of directly controlling flows of transmission lines on which they are located.

8.8 Summary

Many open questions remain relevant for establishing the best role of transmission network-based system regulation in a changing industry. The transmission network is still viewed by and large as playing a monopolistic role, which is in sharp contrast to the generation-based systems control for a competitive market.

There may be a (very long) transition between the present industry stage and the future, when transmission network-based services could be made competitive, e.g., in the sense that individual transmission lines could be assigned economic value. Only the rudiments of this thinking are found in present literature [144]. The major economic incentive to make transmission services competitive comes from the recognition that the charge for generation-based control borne by the need to eliminate transmission line constraints, such as thermal limits, could be eliminated by enhancing critical transmission paths. Thus, generation-based control and transmission are potential gross-substitutes in the economic sense [56]. Much more fundamental work is needed to understand how to implement good economic incentives for transmission service compensation that would make this idea implementable. Instead, the transmission grid for the time being must be viewed as a natural monopoly with guaranteed cost recovery [57].

Given the monopolistic nature of transmission services, the emphasis in this chapter is primarily on technical challenges for designing and operating transmission-based system regulation according to prespecified performance. These challenges are seen in the need to coordinate controls of switching devices on the grid. An interesting opportunity is seen through the use of fast FACTS devices, including the ones which *directly* regulate line flows in specific transmission lines. For the first time the electric power industry has regulating devices one could think of as being *electric valves*. A very hard question in this area, potentially relevant in the near future, concerns coordination of slow mechanical switching of the grid devices, and generation-based systems control over mid- and long-term horizons. Vertical disintegration of the system—its transmission part having a separate owner than the generators could create new dynamic problems over long-term horizons, imposed by mechanical switching and generator-based regulation and their interplay.

The majority of accounting methods for recovering fixed, operating and variable costs associated with the grid are based on short-run marginal cost (SRMC) signals [57, 145] and are incapable of providing incentives for enhancing the transmission grid and its regulation devices. At present, this is an open research problem, but much of the successful technology implementation hinges on providing the right economic incentives.

CHAPTER 9
CONCLUSIONS

The electric power industry in the United States is in the midst of developing responses to the Federal Energy Regulatory Commission's mandate to provide "open access of the transmission system." Being debated are the benefits and consequences of alternative financial, regulatory, and operational designs for unbundling and restructuring the industry. All the alternatives are designed to meet three goals:

1. Providing open access to transmission facilities.

2. Creating competition among generation resources.

3. Preserving system reliability.

The differences among the alternatives are in details of how to meet these goals. It is a question of where and how to draw the lines between the infrastructure and the systems control (ancillary) services required to support that infrastructure. Of particular note is the debate between proponents of a "poolco"[1] and proponents of bilateral scheduling[2] of transactions. Both schemes incorporate the use of a third-party independent system operator (ISO). The ISO is required to maintain system security and reliability in a nondiscriminatory fashion. Much of this debate has taken place on a philosophical level, more concerned with competitive paradigms and analogies to other industries than with informed engineering discussion. Often missing are discussions concerning

- The magnitude and characteristics of the resources needed to keep the system together, i..e., how to preserve system frequency, reserves, and voltages within security limits during normal and emergency conditions.

---
[1] The "poolco" features transactions scheduling effected by a central operator.
[2] The bilateral structure focuses on a decentralized scheme where transactions are effected between individual buyers and sellers.

- The type, amount, and periodicities of information that will be "used and useful" to those interested in participating in network management and protection functions.

- The analytic and software tools available for operational control, planning, and accounting to actually measure the current and expected state of the system.

- The protocols and authority that the ISO must be able to impose under normal and emergency conditions.

We recognize the fluid situation surrounding the industry restructuring. We also expect that from among the wide variety of proposed options, the final contours of the industry will evolve. We are neither attempting to foreclose any options nor asserting what will be the final structure. We are attempting to direct attention to certain crucial problems posed by the laws of physics and by the practical limits on how much information about the state of the system can be calculated and conveyed to the competitive market. These problems must be resolved in arriving at a viable and functional model for a restructured industry.

For purposes of discussion, a framework is imposed on a bilateral energy market overseen by an ISO. The ISO is unaffiliated with any generation, distribution, or market entities. The authors assert that the problems being addressed are somewhat more difficult to handle in the bilateral context than in a " poolco" system. This assertion is based on the fact that the procedures for managing a "poolco"are similar to the planning and control procedures currently applied by "tight" power pools, whereas a bilateral-based system requires determining the extent to which the individually scheduled transactions can be separated from the grid management itself. Other issues in the bilateral environment include the methodology used by the ISO to address (prevent or relieve) flows that constrain those bilateral transactions; and the allocation of responsibility for services (e.g., network protection, constraint relief) among those resources that cause and those that benefit from the conditions requiring those services. We contend that a two-tier market can evolve such that

- Competitive market interchange transactions between third parties are decoupled from network considerations, except for willingness to pay system management fees to the ISO.

- The two markets—a decentralized competitive market (CM) and a centrally directed services market (SM)—can be linked by the ISO in the process of procuring required services from appropriate participants. These services would relate to the state of the network (both actual and projected). The participants in this network services market must be qualified to meet stringent requirements related to availability, controllability, rate of response characteristics, and so on.

The linkage between the two markets is achieved with an iterative process whereby the ISO collects information on transactions, assesses their feasibility, computes service requirements, and supplies the participants in the CM with an SM charge allowing them to continue their market activities.

9.1 Summary of our approach to linking technical and economic processes under competition

Our approach in this text is motivated by the need to quantify relations between the strictly technical, cost-based efficiency that is defined by the existing regulated industry to be a measure of optimal system operation, on the one hand, and the efficiency measures of a competitive industry, on the other.

Utilities, as providers of services to the competitive market, must understand the implications of competitive economics for efficient technical management and pricing of generation-based systems control. Under fairly mild conditions, the generation/demand market may settle close to the economic equilibrium at which the players operate with very little profit. But this does not define the economics of generation-based systems control, which is viewed by most economists as an externality to supply/demand whose cost of service is assumed to be relatively small. No explicit theory exists for fair cost-based compensation, much less for competitive compensation, for these ancillary services. Yet the obligation to serve remains, and the utilities could be exposed to harsh expenses unless rules are well defined.

Two important conclusions emerge, which are particularly relevant for determining the value of generation-based systems control that facilitates competitive supply/demand transactions. First, prior to accounting for the costs of keeping the system together, the theoretical conditions for achieving systemwide efficiency under centralized management are equivalent to the conditions for maximizing the efficiency of each market participant under perfect competition.[3] It is therefore clear that even before the industry moves into further challenges, it is necessary to use costing mechanisms based on cost computations using optimal dispatch instead of split saving formulae and like schemes based on cost averaging over long time horizons.

Second, we suggest that systemwide efficiency under competition strongly depends on efficient pricing mechanisms for generation-based systems control, which plays a fundamental role in facilitating the competitive market. Major obstacles to competitive transactions are seen in transmission congestion as well as potential deterioration in system frequency and voltage. The effective value of a particular market participant level is greatly dependent on the

---

[3]Planning is a separate matter, not discussed in this text.

generation-based systems control cost at which these problems are eliminated.

However, creating meaningful price signals that minimize this additional cost is a complex theoretical problem. Performance criteria essential for keeping the system together are a function of systemwide operating conditions. From an economic viewpoint, they are a common good to all market participants. We suggest that general pricing methods for this type of service are most efficient when pricing is done in a cooperative rather than a competitive mode among the providers [56]. The information balance method (IBM) principle is a particularly useful way of looking at the pricing of ancillary services because it provides the information flow necessary for implementing the suggested pricing mechanism.

Using general conclusions from theoretical economics, we propose a mechanism for defining cost-based prices for generation-based systems control and a possible market implementation. In principle, the concept allows both centralized and bilateral supply/demand transactions. An additional coordinating signal is defined in a centralized manner, and it is used for iterative bidding, which is claimed to approach ideal efficiency. Market structure is open to changing membership in ancillary generation.

The information flow of power quantities into the electric power system (their location and amount) is required at the secondary (coordinating) level, but the prices from market participants are not needed at the coordinating level. It is left up to market participants to respond to the cost of keeping the system together, of which they are informed by the coordinating level. They use their distributed, profit-based criteria to iteratively vary power quantities traded. This framework is an application of the IBM, known for generating price feedback in hierarchical systems when individual performance objectives need to be respected while operating in terms of systemwide efficiency. This method is used to demonstrate the possibility of providing economic incentives to market participants to operate in a manner that reduces the need for systemwide functions or, alternatively, charging them in a nondiscriminatory manner.

The subtle distinction between competitive generation and ancillary generation can be fairly confusing. It is suggested here that a particular supplier of energy has the option to become a provider of generation-based systems control and be paid for it, or just remain a competitive supplier for profit and be charged for generation-based systems control provided by others. Once a supplier is a member of the generation-based systems control set, it becomes a provider of a common good. Membership in generation-based systems control can be made time-varying and therefore open to competitive decision making. If at times it seems that participating in generation-based systems control is less profitable than competing on the market and paying others for ancillary generation, the supplier can switch allegiance at will.

9.2 Relevance of our proposed modeling and control framework

Our proposed modeling and control framework provides a pragmatic means of maintaining reliable system operations while simultaneously fostering a competitive supply/demand market. Implementation details are presented and discussed.

We suggest that the nested hierarchy may be a necessary evolution of the horizontal hierarchy found at present in typical large electric power systems. We propose interaction variables among various levels of a given nested hierarchy as fundamental to deriving minimum-order dynamic models relevant for defining the interplay among subsystems over various time horizons. Control objectives at the interconnected (highest) system level are expressed in terms of interaction variables and participants in systemwide controls only. Our approach allows for control structures that are geared to meeting performance criteria at the level of interest while allowing the interactions with the other subsystems to be uncontrolled. Control flexibilities of this type are necessary for accommodating the needs of a competitive industry, in which different participants are free to specify quality of service desired at a certain price.

The modeling and control approach developed here allows for a unified formulation of generation management in response to competitive transactions while still keeping the system together over mid- and long-term horizons as the open market dictates changes in system inputs. With the framework suggested, it becomes straightforward to define dynamic cost models associated with generation-based controls. Systemwide efficiency in the face of a trade-off between coordination and competition is studied as one of the basic questions in assessing how much competitive markets actually improve systemwide efficiency in electric power systems.

It is emphasized that system operations is more than a series of static optimizations. Operations requires a firm and known base of system resources for its activities. This base is currently provided by operations planners using unit commitment algorithms. This base ensures that sufficient capacity is available to respond to disturbances, sufficient operating bands (between minimum and maximum units' operating limits) to respond to expected load changes, and sufficient responsive generation to respond to short-term load generation mismatches. The proposed framework provides this same base for reliability in the context of an unbundled open access environment. To effect this reliable competitive system the framework differentiates between a decentralized competitive profit-driven market (CM) and a centrally directed services market (SM).

It recognizes the value of incentives, identifies such incentives, and shows how those incentives will bring the two markets together into a *reliable* free market.

9.3 A Final Word

The approach proposed in this text shows that it is critical to differentiate between a systems control (ancillary) services market and a competitive market. The services market is defined as those resources committed (both in the scheduling sense and in the contractual sense) to the reliable operation of the electric power system. The competitive market comprises those resources that function in relation to ad hoc profit motives. In the sense that entities are free to offer their services to either or both of these markets, the proposed markets are both competitive markets. However, the services market resources *must* be committed to *in advance*. The framework's combination of separate markets, a priori commitment and performance incentives integrates the competition of open access with the reliability of the current power system.

It is suggested that under open access the unbundling of resources into CM and SM is essential for managing the system in a reliable manner. In this sense, the concepts described here are different from the "poolco" concepts, which usually suggest active scheduling of all generation. However, a pragmatic look at the potentially huge variety of state variables and response characteristics that must be coordinated by a "poolco" suggests that such scheduling of all functions may not be manageable. Timely response by the ISO to shortfalls in one or more given functions may not be either practical or possible.

In the proposed system the ISO evolves as a nonprofit coordinator of the competitive market and the service market. The ISO effects the following functions:

- Collects state variable information (i.e., loads and generation).

- Projects the feasibility and reliability of the anticipated CM.

- Contracts for needed services from the CM participants who also wish to be SM participants.

- Coordinates services from the SMP participants.

- Monitors real-time activities of the CM.

- Coordinates real-time activities of the SM.

- Allocates SM charges to CM participants and reimburses SM participants.

This framework provides a straightforward integration of a competitive open access market into the real-time operation of the interconnected power system. Both the SM and the CM are made competitive. The qualitative difference is that the resources (they may be generation or load resources) participating in the system services market are scheduled and compensated in a coordinated manner.

Open access raises new issues associated with the ISO. The issues include the establishment of performance bounds used to regulate the system for anticipated requirements and the magnitude of resources that are committed to system services. That commitment today is biased by a system operator's (SO) level of confidence and sense of control of the system. Will this confidence be reduced in the future or at least during the transition? The performance parameters used by the SO today will also change during the transition. Today, the load forecasting errors and capacity reliability data are understood. Decentralized suppliers and loads have no such statistics. Unit characteristics may be drastically altered by how they are utilized. Can the ISO, indeed should the ISO, continue to use these data, or must the ISO look to new statistical evaluators? The ISO will be dependent on the SM to maintain reliability. It is for this reason that severe financial penalties should be imposed for unreliable services by resources committed to the SM. On the other hand, there must be opportunities for rewards for providing and administering these services. The CM participants that are not aware of these opportunities may end up paying more for the services provided by the ISO than they would have incurred providing these services themselves.

There is a peculiar irony in this paradigm for achieving a competitive new world. Rather than focusing on the competitive resources of the market to achieve economies, the open access world may end up focusing on the efficient systems control management and information from the ISO to induce those economies. Only if the ISO's management of the resources participating in the SM is conducted in an efficient and reliable manner will society reap the rewards of restructuring the industry.

BIBLIOGRAPHY

[1] N. Cohn, "Area-wide generation control—a new method for interconnected systems", Electric Light and Power, vol. 31, no. 7, pp. 167-175, 1953.

[2] J. Carpentier, " 'To be or not to be modern' that is the question for automatic generation control (point of view of a utility engineer)", International Journal on Electric Power & Energy Systems, vol. 7, no. 2, April 1985.

[3] M. Ćalović, "Linear regulator design for a load and frequency control", IEEE Trans. on Power Apparatus and Systems, vol. PAS-91, 1972.

[4] J. Zaborszky, T. Chiang, " Economic areawise frequency control," Report SSM 7402, Parts I and II, Dept. of Systems Science and Mathematics, Washington University, St. Louis, Mo., 1974.

[5] L.S. Van Slyck, N. Jaleeli, W. R. Kelley, "A comprehensive shakedown of an automatic generation control process", IEEE Trans. on Power Systems, vol. 4, no. 2, pp. 771-781, May 1989.

[6] D. Ewart, "Automatic generation control: Performance under normal conditions", Proc. of the Systems Engineering for Power: Status and Prospects, Henniker, NH, August 1975.

[7] L. H. Fink, "Concerning power systems control structures", ISA Advances in Instrumentation, vol. 26, Part 1, pp. 1/1-11, 1971.

[8] N. Cohn, "Research opportunities in the control of bulk power and energy transfers on interconnected systems", EPRI EL-377-SR, Special Report on Power System Planning and Operations: Future Problems and Research Needs, pp. 6-1 to 6-30, Feb. 1977.

[9] L. H. Fink (Editor), "Bulk Power System Voltage Phenomena: Voltage Stability and Security", EPRI Report EL-6183, Potosi, MO, Jan. 1989.

[10] L. H. Fink (Editor), "Bulk Power System Voltage Phenomena III: Voltage Stability and Security", Davos, Switzerland, August 1994.

[11] P. Lagonotte, J.C. Sabonnadière, J.Y. Léost, J.P. Paul, "Structural analysis of the electrical system: Application to the secondary voltage control in France", IEEE Transactions on Power Systems, vol. 4, no. 2, pp. 479-486, 1989.

[12] FERC, "Promoting wholesale competition through open access non-discriminatory transmission services"; U.S. Fed. Energy Reg. Comm., 70 FERC 61, 357; p. 175, note 269, 1995.

[13] J. Hamrin, W. Marcus, C. Weinberg, F. Morse, "Affected by the public interest, electric utility restructuring in an era of competition", National Association of Regulatory Utility Commissioners, September 1994.

[14] "Panel Endorses Statewide 'Transco' ", Electrical World, p. 8, August 1995.

[15] Proceedings of the Seminar on "Practical Consequences of a Changing Electricity Supply Industry", October, 1993, ECC Europe BV, Wilhelminsingel 64, 6221 BK Maastricht.

[16] X. Liu, C. Vialas, M. Ilić, M. Athans, B. Heilbronn, "A new concept of an aggregate model for tertiary control coordination of regional voltages", Proc. of the 11th Power Systems Computation and Control Conference, Avignon, France, August 1993.

[17] X.S. Liu, "Structural Modeling and Hierarchical Control of Large Scale Electric Power Systems", Ph.D. Thesis, Dept. of Mechanical Engineering, M.I.T., May 1994 (Thesis advisor: M. Ilić).

[18] M. D. Ilić, S. X. Liu, "A modeling and control framework for operating large-scale electric power systems under present and newly evolving competitive industry structures", International Journal on Mathematical Problems in Engineering, vol. 1, no. 4, October 1995.

[19] " The future of high-voltage transmission", Proc. of FACTS Conference I, EPRI TR-100504, March 1992.

[20] Proc. of FACTS Conference II, EPRI TR-101784, Decemeber 1992.

[21] Notice of Proposed Rulemaking and Supplemental Notice of Proposed Rulemaking (NOPR), United States of America Federal Energy Regulatory Commission, March 1995, Docket No. RM95-8-000.

[22] "Economic valuation of solar/renewable technologies", Third session of the United Nations Commission on Sustainable Development, NY, April 1995.

[23] M. D. Ilić (Guest Editor), Special Issue on FACTS, International Journal on Electric Power & Energy Systems, vol. 17, no. 3, 1995.

Bibliography

[24] J. W. Chapman, "Power system control for large-signal stability: Security, robustness and transient energy", Ph.D. Thesis, Department of Electrical Engineering and Computer Science, M.I.T., February 1996.

[25] A.R. Bergen, Power System Analysis, Prentice-Hall, 1986.

[26] G. Gross, "The new electricity system in England and Wales", a presentation made at the 1993 IEEE PES Triannual meeting, San Diego, CA.

[27] W. Hogan, "Contract networks for electric power transmission", Journal of Regulatory Economics, vol. 4, pp. 211-242, September 1992.

[28] M.A. Crew, P.R. Kleindorfer, "Peak load pricing with diverse technology", Bell J. Economics, vol. 7, pp. 207-230, 1976.

[29] M.D. Ilić, F.C. Graves, "Optimal use of ancillary generation under open access and its possible implementation", MIT LEES TR 95-006, 1995.

[30] D. Bertsekas, Dynamic programming and optimal control, Athena Scientific, Belmont, MA, 1995.

[31] F.C. Schweppe, "Unit commitment scheduling of electric power systems", Proc. of the Systems Engineering for Power: Status and Prospects, Henniker, NH, pp. 116-128, August 1975.

[32] Private correspondence with P. Varaiya, University of California, Berkeley, April 1995.

[33] T. W. Gedra, "Optional forward contracts for electric power markets", Ph.D. Thesis, University of California, Berkeley, 1991.

[34] L.H. Fink, "Impact of electric utility restructuring on energy management and generation control", APPA Engineering and Operations Workshop, Philadelphia, PA, October 1995.

[35] F.F. Wu, P. Varaiya, "Coordinated multilateral trades for electric power networks—theory and implementation", Report PWR-031, University of California, Berkeley, June 1995.

[36] North American Reliability Council, Reliability Assessment, 1992-2001, pp. 18-22.

[37] A. Zobian, M. Ilić, "A steady-state voltage monitoring and control algorithm using localized least squares minimization of load voltage deviations", IEEE Trans. on Power Systems, 1996.(to appear)

[38] F. Wu, P. Varaiya, P. Spiller, S. Oren, "Folk theorems on transmission access: proofs and counterexamples," Internal Technical Report PWR-023, University of California, Berkeley, October 1994.

[39] M.D. Mesarović, D. Macko, Y. Takahara, Theory of Hierarchical, Multilevel Systems, Academic Press, 1970.

[40] W. Findeisen, F.N. Bailey, M. Brdyś, K. Malinowski, P. Tatjewski, A. Woźniak, Control and Coordination in Hierarchical Systems, International Series on Applied Systems Analysis, John Wiley & Sons, 1980.

[41] S. Talukdar, F. Wu, "Computer-aided dispatch for electric power systems", Proc. of the IEEE, vol. 69, no. 10, October 1981.

[42] T. Elacqua, S.L. Corey, "Security constrained dispatch at the New York Power Pool", IEEE paper no. 82 WM 084-2, 1982.

[43] H.W. Dommel, W.F. Tinney, "Optimal power flow solutions", IEEE Trans. on Power Apparatus and Systems, vol. PAS-87, no. 10, Oct. 1968.

[44] J. Carpentier, "Multimethod optimal power flow at EDF", IFAC Int'l Symp. on Power Systems and Plants, Seoul, Korea, August 1989.

[45] M. Ilić, M. Ćalović, "Multicriteria approach to optimal operation of power systems", Journal on Electric Power Systems Research, 1979.

[46] M. Ilić, X. Liu, G. Leung, M. Athans, "Improved secondary and new tertiary voltage control", IEEE Trans. on Power Systems, November 1995.

[47] M. Hasler, C. Wang, M. Ilić, A. Zobian, Margins in power systems using monotonicity", Proc. of the IEEE International Symp. on Circuits and Systems, May 1993.

[48] M.D. Himmelblau, Applied Non-linear Programming, McGraw Hill, 1972.

[49] C.J. Frank, "System operation at optimal cost", EPRI Journal, vol.2, no.9, pp.14-18, November 1977.

[50] M. Ilić, Y.T. Yoon, A. Zobian, "Available transmission capacity and its value under open access", MIT LEES TR95-007, 1995, IEEE Winter Power Meeting, paper no. WM95-295, January 1995.

[51] F. Schweppe, R. Tabors, J. Kirtley, H. Outhed, F. Pickel, A. Cox, "Homeostatic utility control", IEEE Trans. on Power Apparatus and Systems, PAS-99, no.3, pp. 1151-63, May-June 1980.

[52] T. Negishi, "Stability of a competitive economy: A survey article", Econometrica, pp. 635-669, October 1962.

[53] C. Stalon, E.C. Woychik, "Status and direction of California's electric industry restructuring: Summary Report of the Debate in the Competitive Power market Working Group", February 1995.

[54] G. Faulhaber, S. Levinson, "Subsidy-free prices and anonymous equity", The American Economic Review, pp. 1083-1091, December 1981.

[55] G. Faulbaher, "Cross-subsidization: pricing in public enterprises", ibid., pp. 966-977.

[56] J-J. Laffont, J. Tirole, A Theory of Incentives in Procurement and Regulation, The MIT Press, 1993.

[57] A. Zobian, "A Systematic Framework for Cost-based Pricing of Transmission and Ancillary Services in Competitive Electric Markets, Ph.D. Thesis, EECS Dept., M.I.T., August 1995.

[58] A. Zobian, "A framework for cost-based pricing of transmission and ancillary services in competitive electric power markets", Proc. of the American Power Conf., Chicago, IL, April 1995.

[59] M.D. Ilić, F.C. Graves, L.H. Fink, A. M. DiCaprio, "A framework for operations in competitive open access environment", The Electricity Journal, March 1996 (also M.I.T. EL TR95-003, 1995).

[60] J.C. Chow, R. Fischl, H. Yan, "On the evaluation of voltage collapse criteria", IEEE Transactions on Power Systems, vol. 5, pp. 612-620, May 1990.

[61] L. Franchi, M. Iumorta, P. Marannino, C. Sabelli,"Evaluation of economy and/or security oriented objective functions for reactive scheduling in large scale systems", IEEE Transactions on Power Apparatus and Systems, vol. PAS-102, pp. 3481-3488, October 1983.

[62] M.D. Ilić, X. Liu, Ch. Vialas, "Some optimality notions of voltage profile for the steady-state operation of electric power systems", "Bulk Power System Voltage Phenomena III: Voltage Stability and Security", Davos, Switzerland, August 1994.

[63] A. Carpasso, E. Mariani, C. Sabelli, "On the objective functions for reactive power optimization", IEEE Winter Power Meeting, 1980, paper no. A 80 090-1.

[64] E. Kahn, R. Baldick, "Reactive power is a cheap constraint", The Energy Journal, vol. 15, pp. 191-201, 1994.

[65] M. Ilić, J. Thorp, M. Spong, "Localized response performance of the decoupled Q-V network", IEEE Trans. on Circuits and Systems, vol. 33, pp. 316-328, March 1986.

[66] M. Ilić, J. Zaborszky, "Fundamentals of reactive power modeling and control", IEEE publication vol. 87EH0262-6-PWR pp. 61-114, July 1987.

[67] D. Siljak, Large-Scale Dynamic Systems, Elsevier North-Holland, New York, 1978.

[68] D. Hill, I. Marcels, "Stability theory for differential/algebraic systems with applications to power systems", IEEE Trans. on Circuits and Systems, vol. CAS-37, no. 11, pp. 1416-1423, November 1990.

[69] J. Chow, Time-Scale Modeling of Dynamic Networks with Applications to Power Systems, Springer-Verlag, 1982.

[70] J. F. Nash, Jr., "Equilibrium points in N-person games", Proc. Nat. Acad. Sci. U. S., vol. 36, 1950.

[71] J. B. Cruz, Jr., " Survey of Nash and Stackelberg equilibrium strategies in dynamic games", Ann. Econ. Soc. Meas., vol. 4, pp 339-344, 1975.

[72] T. Basar, "On the uniqueness of the Nash solution in linear quadratic differential games", Int. J. Game Theory, vol. 5, pp 65-90, 1976.

[73] M.D. Ilić, Modern approaches to modeling and control of electric power systems, Academic Press, Inc., Series on Control and Dynamic systems, vol. 41, pp. 1-78, 1991.

[74] M.D. Ilić, X.S. Liu, B. D. Eidson, M. Athans, "A new structure-based approach to the modeling and control of electric power systems in response to slow load variations", IFAC Automatica, 1996. (to appear)

[75] M.D. Ilić, "Performance-based value of transmission services for competitive energy management", Proc. of the 26th North American Symp. (NAPS), Manhattan, KS, September 26-27, 1994.

[76] F.C. Graves, P. Carpenter, M.D. Ilić, A. Zobian, "Pricing of electricity network services to preserve network security and quality of frequency under transmission access", Response to the Federal Energy Regulatory Commission's Request for Comments in Its Notice of Technical Conference Docket No. RM93-19-000.

[77] B.D. Eidson, M.D. Ilić, "AGC: Technical enhancements, costs, and responses to market-driven demand", Proceedings of the American Power Conference, Chicago, Il, April 1995.

[78] B.D.Eidson, M.Ilić, "Advanced generation control with economic dispatch", Proc. ofthe IEEE Conf. on Decision and Control, New Orleans, December 1995.

[79] M. Klein, G. Rogers, P. Kundur, "A fundamental study of inter-area oscillations in power systems", IEEE Trans. on Power Systems, vol. 6, no. 3, August 1991.

[80] NYPP-OH TSWG Report, Phase V, Planning Studies, vol. 1, August 1992.

Bibliography

[81] P. Kokotović, H. Khalil (Editors), Singular Perturbations in Systems and Control, IEEE Press, 1986.

[82] M. Vidyasagar, Input-Output Analysis of Large-Scale Interconnected Systems, Springer-Verlag, 1981.

[83] I. Perez-Arriaga, G. Verghese, F. Pagola, J. Sancha, F. Schweppe, "Developments in selective modal analysis of small-signal stability in electric power systems", Automatica, vol. 26, no. 2, 1990.

[84] M. Ćalović, Dynamic State-Space Models of Electric Power Systems, University of Illinois at Champaign, 1971.

[85] J. Zaborszky, J. Rittenhouse, Electric Power Transmission, The Rensselaer Bookstore, Troy, N.Y., 1969.

[86] P. Sauer, M. Pai, "Power system steady-state stability and the load-flow Jacobian", IEEE Trans. on Power Apparatus and Systems, vol. 5, no. 4, November 1990.

[87] C. Desoer, E. Kuh, Basic Circuit Theory, McGraw-Hill Book Co., 1969.

[88] M. Ilić, X. Liu, "A simple structural approach to modeling and analysis of the inter-area dynamics of the large electric power systems, Part I: linearized models of frequency dynamics", Proc. of the North American Power Conference, Washington D.C., October 1993.

[89] M. Ilic, X. Liu, "A simple structural approach to modeling and analysis of the inter-area dynamics of large electric power systems, Part II: nonlinear models of frequency and voltage dynamics", Proc. of the North American Power Conference, Washington D.C., October 1993.

[90] T. Kailath, Linear Systems, Englewood Cliffs, NJ, Prentice-Hall, 1980.

[91] C. Concordia, L. Kirchmayer, "Tie-line power and frequency control of electric power systems", Trans. of the American Institute of Electrical Engineers, vol. 73, June 1953.

[92] Impact of Governors Response Changes on the Security of North American Interconnections, EPRI TR-101080, October 1992.

[93] Proceedings: Automatic Generation Control—Research Priorities, EPRI TR-100451, April 1992.

[94] Transfer Capability, A Reference Document by the North American Reliability Council, October 1980.

[95] W. Stadlin, D. Fletcher, "Voltage versus reactive current model for dispatch and control," IEEE Trans. on Power Apparatus and Systems, vol. PAS-101, no. 10, October 1982.

[96] P.W. Sauer, M.A. Pai, Modeling and Simulation of Multimachine Power System Dynamics, Series on Recent Advances in Control, Academic Press, 1993.

[97] H.G. Kwatny, A.K. Pasrija, L.Y. Bahar, "Static bifurcations in electric power systems: loss of steady-state stability and voltage collapse", IEEE Trans. on Circuits and Systems, vol. CAS-33, no. 10, October 1986.

[98] V. Venkatasubramanian, H. Schattler, J. Zaborszky, "A stability theory of the large differential algebraic systems—a taxonomy", Report SSM 9201, Part I, Dept. of Systems Science and Mathematics, Washington University, St. Louis, Mo., 1993.

[99] M. Ilić, X. Liu, J. Chapman, "Control of the inter-area dynamics using FACTS technologies in large electric power systems," Proc. of the IEEE Conference on Decision and Control, San Antonio, Texas, December 1993.

[100] J.C. Doyle, K. Glover, P.P. Khargonekar, B.A. Francis, "State space solutions to standard $\mathcal{H}_2$ and $\mathcal{H}_\infty$ control problems," IEEE Trans. on Automatic Control, vol. 34, no. 8, 1989.

[101] V. Arcidiono, S. Corsi, A. Natale, C. Raffaeli, V. Menditto, "New developments in the application of ENEL transmission system voltage and reactive power generation", CIGRE Paper No. 38/39-06, Paris, August 1990.

[102] J. Carpentier, "Multimethod optimal power flow at EDF, " IFAC International Symposium on Power Systems and Plants, Seoul, Korea, August 1989.

[103] J.P. Paul, J.Y. Léost, J.M. Tesseron, "Survey of the secondary voltage control in France", IEEE Trans. on Power Systems, vol. PWRS-2, no. 2 , 1987.

[104] J. P. Paul, J. Y. Léost,"Improvements of the secondary voltage control in France", IFAC Symposium on Power Systems, Beijing, 1986.

[105] S. Mondié, G. Nérin, Y. Harmand, A. Titli, "Decentralized voltage and reactive power optimization using decomposition-coordination methods," Proc. of the 10th Conference on Power Systems Computation and Control (PSCC), Graz, Austria, August 1990.

[106] V.I. Idelchik, et al., "Development of methods and procedures to allocate transmission system losses among the parties responsible for such losses" Presented at 11th PSCC, Avignon, France, August 1993.

[107] J.S. Thorp, M. Ilić, M. Varghese, "Conditions for solution existence and localized response in the reactive power network", International Journal on Electric Power and Energy Systems, pp. 66-74, April 1986.

[108] J. S. Thorp, M. Ilić, M. Varghese, "An optimal secondary voltage-Var control technique", IFAC Automatica, April 1986.

[109] X. Liu, M. Ilić, M. Athans, C. Vialas, B. Heilbronn, " A new concept of an aggregate model for tertiary control coordination of regional voltage controllers", Proc. of the IEEE Conference on Decision and Control, Tucson, AZ, December 1992.

[110] M. Ilić, A. Stanković, "Voltage problems on transmission networks subject to unusual power flow patterns", IEEE Trans. on Power Apparatus and Systems, vol. 6, pp. 339-348, February 1991.

[111] M.D. Ilić, F.C. Graves, "Reply comments to the Federal Energy Regulatory Commission's Inquiry on Promoting Wholesale Competition Through Open Access Non-discriminatory Transmission Services by Public Utilities," Docket No. RM95-7-001, October 1995.

[112] S. L. Walton, A discussion paper on forming a single independent transmission system in the Western United States, SL W:, August 23, 1995.

[113] Private communication with the members of the General Agreement on Parallel Path Working Group, 1994-1995.

[114] N. Rao, M. Ilić, "Pricing frequency control in a deregulated electricity market—An application of the interaction balance method", Proceedings of the North American Power Symposium, 63-73, Bozeman, MO, October 1995.

[115] C. Fernando, P. Kleindorfer, R.D. Tabors, F. Pickel, S.J. Robinson, "Unbundling the US electric power industry: A blue print for change", March 1995.

[116] I.B. Rhodes, "Optimal Control of a Dynamic System by two Controllers with Conflicting Objectives", Ph.D. Thesis, Stanford University, 1968.

[117] J.P. Clerfeuille, P. Sandrin, P. Valentin, "Considerations on pricing of transmisison services", UNIPEDE Meeting, Tunis, May 1993.

[118] C-W. Tan, P. Varaiya, "Interruptible electric power service contracts", Journal of Economic Dynamics and Control, vol. 17, pp. 495-517, 1993.

[119] M. Baughman, S. Siddiqi, "Real-time pricing of reactive power: theory and case results", IEEE Trans. on Power Systems, vol. 6, pp. 23-29, February 1991.

[120] M. Baughman, S. Siddiqi, J. Zarnikan, "Integrating transmission into IRP", Parts I and II, 95 WM 200-6 PWRS, 95 WM 201-4 PWRS, IEEE Winter Power Meeting, NY, February 1995.

[121] M. Yehia, R. Chedid, M. Ilić, A. Zobian, R. Tabors, J. Lacalle-Mellero, " A global planning methodology for uncertain environments: Application to the Lebanese power system", IEEE Trans. paper no. 94-WM 224-6 PWRS, 1995.

[122] S. Talukdar, C.W. Gellings eds., Load Management, IEEE Press, 1986.

[123] "Current issues in operational planning," IEEE/PES Summer Meeting, 91SM405-1 PWRS, July 1991, San Diego, California, IEEE Committee Report.

[124] F.C. Schweppe, M.C. Caramanis, R.D. Tabors, R.E. Bohn, Spot Pricing of Electricity, Kluwer Academic Publishers, 1988.

[125] R. Baker, G. Guth, "Control algorithm for a static phase shifting transformer to enhance transient and dynamic stability of large power systems," IEEE Trans. on Power apparatus and Systems, vol. PAS-101, September 1982.

[126] H.E. Brown, Solution of Large Networks, Wiley Interscience Publications, 1985.

[127] R.D. Dunlop, R. Gutman, P.P. Marchenko, "Analytical development of loadability characteristics for EHV and UHV transmission lines", IEEE Trans. on Power apparatus and Systems, vol. PAS-98, 1979.

[128] R. Gutman, "Application of line loadability concepts to operating studies", IEEE Trans. on Power apparatus and Systems, 1988.

[129] R. Reed, J. Willson, "The PJM approach to detect, analyze, and operate the system considering reactive problems", IEEE Tutorial Course, 87#0262-6-PWR, pp. 28-34, 1987.

[130] I.M. Gelfand, S.V. Fomin, Calculus of Variations, Prentice-Hall, 1960.

[131] C.A. Desoer,"The maximum power transfer for n-ports", IEEE Transactions on Circuit Theory, vol. CT-20, May 1973.

[132] C.A. Desoer, E.S. Kuh, Basic Circuit Theory, McGraw-Hill, 1969.

[133] A.J. Calvaer, "On the maximum loading of active linear electric multiports", Proc. of the IEEE, vol. 71, no. 2, February 1983.

[134] F. Spinei, "Determination of the states of maximum power transfer and of power transfer with maximum efficiency in networks with modifiable parameters, Part I: Resistive DC Networks," Rev. Roum. Sci. Techn.-Electrotechn. et Energ., 16, 3, Bucharest, voll. 16, no. 3, 1971.

[135] P. Cristea, F. Spinei, R. Tuduce, "Comments on 'Reciprocity, power dissipation, and the Thevenin circuit", IEEE Trans. on Circuit Theory, October 1987.

[136] M. S. Ćalović, "Modeling and analysis of under load tap changing transformers in voltage and Var control applications", University of Illinois at Urbana-Champaign, Report PAP-TR-83-3, 1983.

[137] J. Medanić, M. Ilić, J,. Christensen, "Discrete models of slow voltage dynamics for under load TAP changing transformer coordination," IEEE Trans. on Power Systems, PWRS-2, pp. 873-882, November 1987.

[138] K. Lim, M. Ilić, "Control coordination on the distribution power network", Proc. of the IEEE International Symposium on Circuits and Systems, Singapore, June 1991.

[139] F.K. Mak, M. Ilić, "Towards most effective control of reactive power reserves in electric machines", Proc. of the 10th Conference on Power Systems Computing and Control Conference, Graz, Austria, August 1990.

[140] J. W. Chapman, M. Ilić, C. King, et al., "Stabilizing a multimachine power system via decentralized feedback linearizing excitation control", 92 SM 540-5 PWRS, IEEE Summer Power Meeting, Seattle, July 1992.

[141] J. Zaborszky, K.W. Whang, K.V. Prasad, "On line stabilization of the large interconnected power system", SIAM Conference on Electric Power Problems: The Mathematical Challenge, Seattle, March 1980.

[142] R. Marino, "On the stabilization of power systems with a reduced number of controls," Lecture Notes in Control and Information Sciences, Springer-Verlag, pp. 259-272, pp. 598-611, 1990.

[143] M. Ilić, X.S. Liu, "Direct control of inter-area dynamics in large systems using flexible AC transmission systems (FACTS) technology", patent, 1995.

[144] H-P. Chao, S. Peck, "Market mechanisms for electric power transmission", IAEE meeting, Washington, D.C., June 1995.

[145] A. Zobian, M. Ilić, "Unbundling of transmission and ancillary services", IEEE paper no. WM96-732, IEEE Winter Power Meeting, Baltimore, MD, January 1996.

[146] M. Ilić, Y.Yoon, A. Zobian, "Available transmission capacity", IEEE paper no. WM96-295, IEEE Winter Power Meeting, Baltimore, MD, January 1996.

[147] M. Ilić, C-N. Yu, "Minimal system regulation and its value in a changing industry", IEEE Summer Power Meeting, 1996.

INDEX

achievable systemwide efficiency 33
allocation of responsibility 18
allowable line flows 39
ancillary generation 33
ancillary services 33
area control error (ACE) 22
automatic generation control (AGC) 18
automatic secondary voltage control (AVC) 32

benefit function 33
benefit maximization 43
bilateral-based system 18
bilateral structure 16
bilateral transactions 50

capacitor
    series 199
capacitors 199
centralized futures markets 50
centralized industry structure 33
charging mechanism
    optimal 170
closed-loop system regulation 20
common good 49, 222
competition 47, 170
competitive energy market 18
competitive market participants (CMPs) 16
congestion cost 39
conservative scheduling 31
constrained economic dispatch (CED) 34
contingencies 31
contract path 40
contracts

    firm 24
    non-firm 24
control
    fringe 109
    generator 216
control area 29
control areas 16
cooperative pricing 33
coordination 47, 170
coordination of technical management 33
coordination over large geographical areas 42
cost-based efficiency 221
cost/benefit optimization 38
cost allocation method 48
cost function 33
cost minimization 45
cost of coordinated ancillary services 48
cost of loss compensation 36
cost of reactive power 39
costs of keeping the system together 33

DC transmission 199
decentralized regulation 50
demand curve 24
deregulated industry 18, 34
discrete event process 141
distributed energy market 44
distributed optimization objectives 33
dynamic control schemes 59
dynamic optimization 21, 31
dynamic programming techniques 20

Index

dynamics of competitive markets 20

economic dispatch/scheduling 36
economic efficiency 33
economic equilibrium 46, 221
efficiency 51
efficiency over long time horizons 41
EHV transmission 24
electric valves 217
emergencies 31
emergency state 26
energy of prespecified quality 19
energy transactions 16
equal treatment 170
externalities 46

flexible AC transmission systems (FACTS) 208
flow scheduling 140
frequency regulation 32
fully distributed structure 16

game for profit 44
generation-based controls 19
generation-based service 20
generation allignment 140
generation excess 42
governor control 198
governors 26
gross-substitute 217

hard constraint 37
homeostatic controls 42
horizontal hierarchy structure 11
hybrid structure 16
hydro storage facilities 26

ideal economic efficiency 44
ideal technical efficiency 33
impedance matrix 199
inadvertent energy exchange 50, 109
inadvertent energy exchange (IEE) 54
incentives for optimal demand 43

independent power producers (IPPs) 14, 42
independent system operator (ISO) 18, 46
indicators of congestion cost 40
industry structure 170
input changes 23
inter-area oscillations 213
interaction variables 22, 121
intra-area dynamics 216

Lagrangian 37
line-flow control
    direct 209
load flow equations 37
load following 30, 36
load frequency control
    automatic generation control 32
load model 51
load power factor 203
long-run incremental cost 41
loss compensation 20, 36
lossminimization 140

manual control 26
market-price modulated demand 14
market players
    competitive 171
maximizing social welfare 43
maximum power transfer 200
    constrained 203
metering
    communications 34
mid- and long-term horizons 32
minimal information 31
minimal regulation 172, 175
minimum-order models 21
most efficient price signals 48
multi-utility environment 42

Nash strategy 154
native load 14
need for coordination 47
nested hierarchy 11
nested hierarchy structure 14

new control structures 18
new monitoring and control structures 18
non-uniform power quality 18
non-utility-owned participants 18
noncompliance 24
noncompliance with the market transactions 20
nondiscriminatory pricing 45
nonperfect market 44
normal operating mode 31
normal state 26

obligation to serve 169, 221
open access system 33
operating reserve 12, 19, 26
optimal compensation 202
optimal dispatch 33
optimal power flow 34
optimal voltage control design 32
optimization objectives 33
out-of-merit cost 41
out-of-merit generation dispatch 12

partially distributed decision making 33
path-by-path congestion cost 40
path provider 35
penalty function 40
perfect information 46
performance criteria 30
performance objectives 23
poolco structure 16
power trade 44
preventive operating mode 31
price-competitive generation and demand 43
price-driven demand 18
price-driven generation 18
prices
    volatile 171
private nature of goods 46
profit 46
profit-driven generation unit 42
profit optimization 44
projecting the market profile 21

random fluctuations in native load 20
reactors 199
reduced information 18
regional transmission group (RTG) 42
regulated industry 33
regulation 26
    minimal 109
    network-based 195
    reactive power/ voltage 119
regulation problem at the interconnected system level 19
reliability 51
required control performance 26
residential load 24
restarting/shutting off 26
retail wheeling 14, 24, 42
revenue maximization 45

security criterion 12, 31
security enhancement 140
short run marginal cost (SRMC) 35
single administrative unit 16
social welfare maximization 33
soft voltage constraints 38
static optimization 21, 31
static Var compensator 208
storage 26
stranded cost 35
switching
    mechanical 199
switching capacitors 36
systems control 23
system services 16
systemwide optimum 46

tight pools 18
time scale
    control-induced 30
total cost 33
total supply/demand mismatch 22
transfer
    real power 200
transfer limit

Index

steady-state 200
transformer
    phase-shifting 204
transmission constraints 39
transmission limits 40
transmission loss 36, 202
two-level hierarchy 46

unconstrained economic dispatch (ED) 34
unit commitment 21, 26
utility functions 47

value
    performance-based 170
vertical disintegration 217
voltage/reactive power support 35
voltage constraints 37
voltage optimality 39
voltage regulation 32

wheeling mode 12